# SAS

## ULTIMATE GUIDE TO COMBAT

D1341142

OSPREY
PUBLISHING

# SAS

## ULTIMATE GUIDE
## TO COMBAT

### HOW TO FIGHT AND SURVIVE
### IN MODERN WARFARE

ROBERT STIRLING

First published in Great Britain in 2012 by Osprey Publishing,
Midland House, West Way, Botley, Oxford, OX2 0PH, UK
44-02 23rd Street, Suite 219, Long Island City, NY 11101, USA
E-mail: info@ospreypublishing.com

OSPREY PUBLISHING IS PART OF THE OSPREY GROUP

A CIP catalogue record for this book is available from the British
Library

ISBN: 978 1 84908 764 3 (UK edition)
ISBN: 978 1 84908 763 6 (US edition)
Ebook ISBN: 978 1 78096 399 0
Epub ISBN: 978 1 78096 400 3

Page layout and all illustrations by Ken Vail Graphic Design
Typeset in Conduit ITC, Geogrotesque Stencil and Haettenschweiler
Originated by PDQ Media, Bungay, UK
Printed in China through Worldprint Ltd

12 13 14 15 16   10 9 8 7 6 5 4 3 2 1

Osprey Publishing is supporting the Woodland Trust, the UK's
leading woodland conservation charity, by funding the dedication
of trees.

www.ospreypublishing.com

## AUTHOR'S NOTE

The author is extremely grateful to Colour Sergeant Trevor
Sadler, formerly of 1st Battalion Royal Anglican Regiment,
Sergeant Roy Mobsby, formerly of 63 Parachute Squadron
RCT, and sergeants Robert Collington and Tom Blakely,
formerly of the Parachute Regiment, for supplying some
of the photographs used in this book. He would also like to
thank Colour Sergeant Michael McIntyre BENM, formerly of
the Parachute Regiment, for his generous assistance and
advice in many areas.

## EDITOR'S NOTE

For ease the following data will help in converting imperial
measurements to metric and vice versa:

1 mile = 1.6km
1lb = 0.45kg
1 yard = 0.9m
1ft = 0.3m
1in = 2.54cm

## DISCLAIMER

# CONTENTS

# CHAPTER 1

# Introduction: Why Do You Need This Book?

This book will teach you a wide range of advanced soldiering skills known only to the most experienced and highly trained soldiers in the world. If you want to understand everything the switched on Special Forces operator needs to know, plus quite a lot more relating to planning military operations, then read on.

The author 2011: some places you need a beard to be taken seriously. (Author's Collection)

You may already be a soldier of some kind and presumably you want to be both a better soldier and to stay alive while you are doing it. The chances are that if you are a soldier then you are either going on operations where some guys come back in body bags or you want to join the particular Special Forces (SF) group that is your idea of the dog's gonads: British Special Air Service (SAS), US Navy SEALs, US Army Rangers, German Jagdkommando or Russian Spetznaz. And the only reason you would join a Special Forces unit would be to go where the action is. Isn't it?

So one way or another you are going to fight. And if you are going to fight as a soldier then you need to know what you are doing because, in the end, you must either win or you will die. For sure, against a force of Taleban or al-Qaeda insurgents, surrender is not a realistic option. Not if you want to die an old man in bed with all your bits attached – so you either succeed in your mission and stay alive or, eventually, you don't. To keep your mind on the job I will just mention here that I have seen insurgents in Afghanistan dig a hole like a deep grave and throw some sticks in the bottom. Then they doused the sticks with petrol and tossed a match in so it lit up like the gateway to Hell. When it was nice and warm they rolled their prisoners in. Just with their hands tied behind their backs so they could jump about a bit but not get out. You do not want to end up in a similar situation.

What going on operations – playing on the two-way range – and joining a Special Forces unit have in common is that you need a thorough knowledge of soldiering skills to stay alive. Knowledge ranging from what the grunts are taught in basic training through to the skills taught to Special Forces operators right up to the wide range of military knowledge required of those tactical officers who plan special operations. This is where you will get that knowledge. Whatever you do as a soldier at the moment, the chances are that in this book you will learn a few military skills and about bits of kit you just never had the chance to come across before.

You might say, quite rightly, that any knowledge only improves your odds of survival on the battlefield. What I would say in reply is that some units in Afghanistan, as I am writing this, are taking 10% casualties killed and seriously wounded. Are you happy with a 90% chance of coming back in one piece or would you rather lift the odds to 99%? It's up to you.

# WHO AM I TO TELL YOU ALL THIS

At this point you might be wondering who the hell I am to tell you all this. Well I'm no Arnold Schwarzenegger or Captain America. I've just done a bit of soldiering here and there and been in a fair few firefights. I was a British Paratrooper, tank soldier, passed SAS selection at 19 and did some covert work in Northern Ireland before leaving the British Army for a bit more action

The author teaching unarmed combat to some interesting characters at the Vienna Military Academy Austria 2009: notice the foot. (Author's Collection)

The Author in Rhodesia, 1979, with part of a .50cal Browning machine gun. You don't want to carry one very far. (Author's Collection)

The Author, back left, with a Special Forces team somewhere in southern Africa, 1982. (Author's Collection)

in the African Bush Wars of the 1980s. I've only been wounded twice and I've learnt from that. I still train European Special Forces in unarmed combat and various units around the world in all sorts of soldiering skills.

You may think that a bit of experience of staying alive, and a lifetime of working with various armies across the world, gives me some grounds to tell other soldiers how to stay alive and do some damage to the enemy. Or you may not – I can stand criticism from anyone who has seen a few rounds fired in anger.

# HELPING THE ENEMY?

Some people have accused me of helping the enemy by providing them with soldiering skills to use against our guys. I tell them not to believe everything they see on TV. Terrorists and insurgents are, for the most part, neither stupid nor badly trained. And anyone who says they are is either ignorant or trying not to frighten the children. Insurgents already know everything in this book that they can use. Insurgents get better training than most infantry soldiers so you can be sure that this book is not helping them. In any event, the roles they play are quite different and the tactics useful to a peacekeeper for staying alive while operating a vehicle checkpoint, for example, are quite useless to the modern terrorist.

# WHAT IS INSIDE?

This book is effectively a training manual to raise your soldiering skill levels to those of operators serving with the best of the world's modern Special Forces, from the US Navy SEALs through the Russian Spetsnaz to the SAS. On top of that we also look at a wide range of weapons and tactics useful to the Special Forces officer in planning operations.

The way I have laid it out is as follows:

- **Personal Weapons** – so you know how to shoot anything you come across
- **Food & Shelter** – so you can keep cool, warm, dry and get yourself something to eat
- **Staying Healthy** – so you can dodge exotic diseases and patch your mates up
- **Not Getting Shot** – so you can come home and collect those shiny medals
- **Not Getting Shelled** – so you can come home in one piece
- **Not Getting Mined** – so soldiering won't get in the way of your dancing skills
- **Dodging Suicide Bombers** – so you don't visit paradise too early yourself
- **Tactics For Defence** – so no one can take that flag off you
- **Tactics For Attack** – so you know how to win the medals

Of course there is a lot to cover here so to make it even more interesting than watching paint dry I have thrown in a few war stories to make you laugh, make you cry and, most importantly, make the ideas stick in your head.

A fair demonstration of effective camouflage, an essential for the modern soldier. (USMC)

# SEAL Operation: Kill Bin Laden

We can learn a number of vital lessons from the successful operation to kill Bin Laden.

Lesson One is the vital importance of intelligence from operators in the field and the interrogation of prisoners. Starting with just a clue in 2002, British and US agents in Pakistan, and US Intelligence personnel interrogating terrorist prisoners, gradually tracked down Bin Laden to a town called Abbottabad, 31 miles north-east of the Pakistani capital, Islamabad. By April 2011 further intelligence efforts confirmed that that there was a 90% chance that Bin Laden was living in a large house at Abbottabad within a one-acre compound surrounded by 18ft walls. Surprisingly, for a high-status dwelling, the house had no telephone or internet connection. This was deemed suspicious. Based on this information, and more, President Obama made the decision to send SEAL Team 6 to kill Bin Laden – a total force of two dozen men and an SF dog named Cairo. The decision was also made not to inform the Pakistani leadership, although they are technically allies, of the planned assault. This meant that the local intelligence efforts as well as the success of the mission were not compromised.

On the night of 2 May 2011 surveillance aircraft blocked communications over a wide area as two secret stealth helicopters – an advance on the Blackhawk/Pavehawk model – brought in the SF troops. All the while fighters and gunships patrolled the skies overhead. The original plan was that one helicopter would drop a team by rope onto the roof of the building and one would land in the compound. However, because of ground effect air blast created by the massive compound walls one of the helicopters became unstable. Even the ace pilot flying it was unable to keep control and it crashed into the compound wall.

Lesson Two therefore is that SF soldiers need to think on their feet – and the plan was hastily revised to enter the buildings from the ground and work upwards. Little armed resistance was met and both Bin Laden and one of his wives were shot dead, the latter as she attempted to protect him.

Lesson Three is that a live prisoner is not worth the risk to the SF operators if there is even the slightest suspicion that the detainee could be armed or wired. But securing the body ('proof of death') was crucial to ensure the success of the operation. As a result the body was flown out of the compound and the necessary DNA evidence secured before burial at sea. SF operators must also ensure that no vital equipment or information falls into the hands of the enemy. Therefore the crashed helicopter was destroyed with explosives.

All personnel returned safely to base. Overall a very well run operation using SF troops who could clearly think on their feet and act independently when required.

# CHAPTER 2

# Tools of the Soldier's Trade: Your Weapons & Personal Equipment

In this section I've covered all the main groups of weapons you are likely to be issued with or come across in your military life; also some personal equipment for navigation and communications as this sort of fits in nicely.

The average terrorist doesn't use the same Quartermaster as us so besides principally British, American and European equipment we will be looking at Russian and Chinese kit as these are the well-armed insurgent's favourite suppliers. Both countries produce the RPG7 rocket launcher and the Kalashnikov assault rifle for instance.

Some readers will already be well trained in most areas of soldiering skills but it is my aim to ensure you all know what you are doing — whether new recruit or Special Forces operator — to a very high standard. This will make you a better soldier and keep you alive longer. To ensure everyone follows along and hasn't missed 'Chapter 1' I will be explaining everything as if you had never seen it before. Bear with me as the more you know the better chance you have of surviving any particular confrontation and most of us have forgotten a thing or two from training.

In any event, knowing the capabilities and potential of particular equipment focuses the mind when it is being used against you and gives you the confidence to plan and respond appropriately. You know what the insurgent, terrorist or guerrilla fighter can do and you know what is myth. It's always better, to my mind, to know the true facts then you can make better decisions based on them. Knowledge dispels fear. Which is always a good thing when everyone around you is being shot to pieces.

Knowing the facts about what you are up against is the first step in defeating the threat.

When all is said and done, the better you know your weapons, and your enemies' weapons, the better you can kill the enemy and prevent him from killing you. And that is the name of the game isn't it? What you will learn here is what guns, rockets, mines and so on can really do — and what they can't do so the movies don't give you the wrong idea.

You will see that, so far as possible, I have covered each weapon under a series of headings. My idea is that this leads you to look at the weapons here, and any similar type of weapon, in a similar way. You will therefore be able to assess any weapon you come across and know what questions to ask when considering it for a combat role and how to compare it with other weapons of its type. Rate of fire, range reliability, that sort of thing. So in each section you will see the following (with just a few exceptions) which in line with a long military tradition we will remember the headings with the mnemonic GHOSTS:

- **Goal:** When is the right time to employ this weapon? What is it good for?
- **History:** How did it develop to be the way it is?
- **Operation:** How does it work?
- **Skill:** How do you use the weapon to maximum efficiency?
- **Types:** Specific examples of this weapon you are likely to come across
- **Summary:** What are the important things to remember?

After all I have said here and elsewhere about clever weapons, I believe that you, the soldier, are the real weapon – you don't need fancy guns to be dangerous you just need the right training and attitude. You can steal everything else. The only thing you can't do without is your boots. So just to upset the obvious order of things we'll start with your boots because if you can't walk and you're out in the sticks with no support then you will die.

# BOOTS AND OTHER FOOTWEAR: MARCH OR DIE!

A civilian opening this book might wonder why I'm starting with a lecture on footwear. Shouldn't I be starting with machine guns and things that go bang to catch your attention? The veteran will smile.

Anyone who has gone so far as getting through basic infantry training will understand that it's not called 'boot camp' for nothing. Even though the infantry soldier nowadays often has transport laid on in the form of helicopters and various types of Armoured Personnel Carriers, much of his time is still spent on foot because the infantry soldiers' primary job is to 'take and hold ground'. You take ground by walking over it and taking it from anyone arguing and you hold ground by either digging in and sitting there or, again, walking over it on patrol. Even with a dug-in static position you do clearance patrols around the position morning and evening.

Tanks, missiles and aircraft cannot hold ground nor take ground. They just make life miserable for whoever has it at the moment. They cannot storm dug-in positions, bayonet everyone inside and hoist the good guys' flag. They cannot creep undetected hundreds of miles inside enemy territory to take photos, shoot individuals or blow things up. Taking ground, holding ground and tabbing* over the mountains is what we 'grunts' do. On foot.

---

* Tabbing comes from the British Airborne acronym Tactical Advance to Battle. It means a long run or march with heavy kit.

And the further up the ladder you go towards being a heroic Special Forces operator the more walking you are going to do and the more important your feet will become. Even with all the clever motorized transport there is nowadays, much of the real work of a Special Forces soldier involves walking towards or away from the enemy carrying a load of kit that would stun a mule. And the more weight on your back, the more pressure on your feet.

In anti-insurgency warfare you will spend many happy hours walking up and down streets showing the flag, through

The Author, left, after a 40-mile march across the Drakensberg mountains with rifle and Bergen under the African sun. The Zulus call these mountains *uKhahlamba* which means 'barrier of spears'. (Author's Collection)

villages and maybe along dirt roads so you can find the mines before the valuable armour gets there. Depending where in the world you are you may be walking through endless desert, bush or jungle. It may be very hot, very cold or wet all the time. You may be a long way from support. Sometimes you will be running. And trust me, being chased by a thousand screaming beards is not a good time to have sore feet.

All this covering ground on foot presupposes you can stand on your feet. If at some point your feet won't carry you, you must hope they can get a chopper to you before the enemy catch up. How many yards can you move without putting one of your feet to the floor? On a covert patrol behind enemy lines a soldier who cannot walk tends to be a dead soldier. He is certainly a useless soldier and may well be a liability for needing others to carry him or heroic chopper crews to come in dodging the missiles and fetch him out.

How many miles can you carry a man who can't walk? What would you do if two of you were 50 miles behind enemy lines, your mate had lost the flesh off the bottom of his feet and there was no one to come and collect you? This is real soldiering and not the movies where they 'leave no man behind'. I wouldn't risk the lives of a patrol by carrying a man and slowing them up to save an idiot who didn't take care of his feet. And shooting a wounded comrade and friend so the enemies' women cannot play with him is not something you want to carry in your head for the rest of your life.

## What do your feet need to stay healthy?

I have had African friends who could run a hundred miles in their bare feet and think nothing of it. We Westerners have grown up wearing shoes to keep our feet warm, clean and looking smart. This has made them soft so in the forces we wear boots to protect our tender feet from sharp rocks, which would cut them. But it is also the case that wearing the right sort of boot protects the sole of the foot from wear so well that, properly set up, we can cover distances which would grind the skin from the toughest African's foot. Of course, if your boots are not set up properly you will get blisters...

But it is not just protection from wear that your feet need. We are not always tabbing across the African savannah so sometimes it is cold and our feet need to be kept warm or we get frostbite and our toes go black and have to be cut off. Sometimes it is wet too, and when your feet get wet the skin, no matter how tough you are, falls off in just a few miles. Even if it doesn't get to the skin falling off, damp skin breeds fungus and this makes it split and bleed. Not nice at all.

Royal Marines trudge through deep snow during exercise *Himalayan Warrior 07*. Learning to deal with adverse weather conditions is a critical component of Special Forces training. (Dave Husbands © UK Crown Copyright, 2007, MOD)

So, all in all, keeping our feet healthy and happy is vital to our staying alive and doing our job as a soldier. To keep them healthy we have to select the right footwear to keep our feet protected, dry and warm. Easy. Actually, depending on where in the world you are, sometimes it is not.

## Feet in temperate conditions

From 20 or 30° below freezing to well over 100° in the shade your feet can get by quite well in leather boots so long as you keep them clean, dry and well powdered.

Good leather boots have tough yet supple uppers and stiff soles to allow comfortable marching and have some degree of waterproofing, and yet they can breath a little to let the sweat out. If you visit any hiking or outdoors shop you will find countless brands of boots which will do this, and more, in a very efficient manner – at a price. There are boots made of materials which will let the sweat out very well, fresh air in, keep the water out and keep your tootsies warm far better than leather but these things cost money and soldiers in the British Army have a lot of kit to buy owing to the issue gear being so generally rubbish.

Having got yourself the best boots you can, and broken them in with lots of marching when a blister doesn't matter, on operations you should wash your feet every day if you have the water, dry them and apply foot powder. Put on a clean pair of socks and wash the spares – if necessary carry them tied to the outside of your pack or sleep with them inside your sleeping bag to dry them.

## Feet in cold conditions

Where the climate is cold you should have boots plenty big enough and wear thermal socks if you can get them. Feet are not really designed for very cold climates, being at the extremity of the body and poorly supplied with blood, so they need extra attention to keep them warm when it is chilly outside. The reason for wearing boots large is that tight boots cut off the blood supply to your feet and encourage them to get cold and die. As you walk the action of your foot changing shape and pressure pumps the blood around your feet, if they are not strangled in your boots, and this blood flow acts like the hot water in a household radiator system. If the blood supply is cut off in a cold climate, or if your feet just get too cold, ice crystals will form inside the flesh. This is frostbite and kills the affected parts permanently. Check regularly for frostbite and you may keep all your toes. Leave it too long between checks and you will lose toes and bigger pieces.

Frostbite starts off with 'frostnip', which is a freezing of the outer skin. When frost-nipped the skin turns white as the circulation reduces then it begins to sting or tingle. As it turns to frostbite as the flesh freezes, the area becomes numb and the skin may turn red then purple-white. People have often said that the area begins to feel wooden. When the affected area has frozen properly the flesh is dead and cannot be revived. Amputation is the only answer and while you are waiting gangrene will set in – the flesh will turn black, stink to high heaven and begin to spread poison around your body. Keep your feet warm.

Frostbite, before the toes were cut off. (Corbis)

## Feet in hot conditions

Where the climate is very warm then sweating encourages the fungus, which we all have permanently on our feet, to overcome the natural balance and become noticeable. At best this itches and at worst it eats your foot away or makes the skin crack and bleed. I have experimented successfully with wearing strong sandals in hot dry climates when sand is not a problem. These work so well because the gaps in the upper allow a free flow of air to keep your feet dry and fairly cool so the fungus doesn't grow, while the strong sole still protects your hard, dry feet very well. Africans often wear 'Thousand Milers' which are a sandal made with a sole cut out of a car tyre. Sandals do work well when there is no sand and they keep your feet dry and healthy, but strong boots give a soldier a feeling of security.

In very sandy conditions wearing sandals allows sand to get between your toes or between your feet and the sole of your footwear and this makes your feet bleed – so sandals are not such a good idea here. Canvas boots work quite well in this situation as they offer some of the ventilation advantages of sandals but keep the sand out. If you can't get them you may have to stick with leather boots or 'Desert Boots' made from suede. In this case keep your feet clean and dry so far as you can with extra washing, sock changes and powder to kill the fungus.

## Feet in wet conditions

The worst thing for your feet is water. Have you ever sat for hours in a bath and seen your feet go all crinkly? Soaked wet feet fall to pieces when you walk on them. The skin flakes off and leaves raw flesh. Dry your feet whenever you can. In wet jungle conditions you may just have to put up with it. And with the leeches that get in through the lace holes. In very wet conditions it may be best to wear canvas boots which dry out quickly but frankly there is no real answer to the permanent wet of, say, a swampy jungle and the only thing to do is keep the time spent there as short as possible. Of course, that is a planning matter and not down to the men at the sharp end.

## Encouraging the men

Except in jungle conditions or extreme cold, bad feet should be considered malingering and self-inflicted wounds. Sometimes it will be a miserable soldier trying to work his ticket home and sometimes it will just be depression de-motivating them. On active service, with some soldiers in some conditions, the mind can 'forget' to deal with personal hygiene tasks while it is either bored stupid or concentrating on other matters. The offender should be punished severely to encourage the others. For everyone's good.

If you don't take care of your feet they will get sore. And you would not be laughing if I had anything to do with it. (US Army)

## Blisters

Where I have spoken above about frostbite and rotting feet I was warning you about what can happen in the extreme conditions you may sometimes have to operate in. What every soldier will have to deal with are blisters. Blisters are bubbles of fluid, which form on any part of the foot where the skin has been rubbed too severely against the boot through the sock. A self-protection mechanism of the body then causes the surface of the skin to lift and form a bubble filled with clear fluid or blood. Blisters are the body's own best effort to stop you doing whatever is causing injury to your foot.

The best way of dealing with blisters is to avoid them. Keep your feet dry so the skin stays strong, wear well-fitting boots with stiff soles, so they don't rub as they flex, and take advice on the best combination of socks available. Technology is progressing in these things like everywhere else and some socks now slide to protect your feet. Above all, harden your feet with plenty of marching during training when sore feet don't matter.

Whatever you wear your feet will rub as you walk. By regular long marches your feet will become harder and blisters rare. If you do plenty of marching in your chosen combination of boots and socks then the boots will mould to fit your feet and your feet will gradually build up thicker layers of skin where there is any friction.

In the old days soldiers used to piss in their boots to make the leather pliable and then wear them laced up tight until they dried so that they moulded to the foot. I have done this myself – and the wimpy alternative of soaking boots overnight in a bucket of water – both of these tricks work really well with new leather boots. Boots fit better nowadays straight from the Quartermaster or boot-shop but the principle of getting boots that fit well still applies.

If, during a long march, you sit down and take your weight off your feet any blisters will fill up with fluid and make the next few miles uncomfortable. If you take your boots off in mid march then, if the blisters are bad, you may not get the boots on again without draining the blisters. In the South African Army we used to do long marches on hard roads carrying a lot of weight – 70lb or more. To deal with blisters we were issued with syringes and a bright red anti-septic fluid based on iodine. When you got a blister you sucked out the fluid with the syringe then injected the red fluid into the sack. This made strong men hop around but it killed the nerves in the area and allowed pain-free walking for a while. Very practical people the South Africans.

If you don't have this facility slice the blisters with a blade, dry the area with powder and cover with a plaster or duct tape. It's not a perfect solution but it will get you through a march with less discomfort than doing nothing. It is a fact that in all areas of life human beings differ. Some people are taller, some fatter and some cleverer. Some people have tougher skin on their feet than others so they get fewer blisters and their feet, once hardened by marching, stay harder for longer. My own feet are pretty soft and once hard go soft quickly so I have to take special care of them. Life just isn't fair so you play your cards the best you can.

I once knew a guy from the French Foreign Legion who would never use soap on his feet as he believed it softened them. He washed his feet in plain water every day. I don't know if this was useful or not but it shows how important a professional soldier considers his feet to be.

Blisters are never, ever an excuse to quit a march. They only sting a little and you should keep going. Certainly no real soldier would ever think of stopping for blisters. And as far as Special Forces units are concerned, a mate of mine ran with pack and rifle for 7 miles after breaking his hip on a rock in the snow. It was on selection for the SAS – the part called the Fan Dance where you run over the mountain Pen y Fan twice lugging a heavy pack – and the staff failed him for being 5 minutes too slow. We don't want wimps in the Special Forces do we?

# RIFLES

## Goal: When is the right time to employ this weapon? What is it good for?

The military rifle is only designed to do one thing: kill other human beings – and it does this both effectively and selectively. The fact that it can be used as a threat to keep the peace may be thought of as a bonus. Although in the past the rifle was the main way to kill the enemy in bulk, today mass destruction is best achieved by air assets or artillery and you use a rifle either to protect yourself, kill the enemy in small numbers or when you cannot use bombs or artillery to destroy a city or camp for fear of harming civilians.

Though you may use different weapons in different conditions, the rifle remains the best way we have of applying concentrated force to kill the bad guys and leave the good guys, and the women and children, alive. The idea is that the infantry soldier, or the Special Forces operator, gets close enough to see an enemy doing something wrong, or being in the wrong place, and, by means of their rifle, terminates their activities permanently. This explanation may seem strange to you. Think about it alongside what you will learn in the rest of this book.

## History: How did it develop to be the way it is?

The ancestor of the rifle was the musket, which began to make an appearance around 500 years ago. This was a huge, heavy, smoothbore weapon employing gunpowder to drive a heavy lead ball down a tube with very modest accuracy. But it would penetrate a wall

The SA80 Assault Rifle – the standard issue to the British Army – and a poor choice in my opinion. (Gaz Faulkner © UK Crown Copyright, 2009, MOD)

never mind a suit of armour and this is what essentially changed warfare away from the knights and horses era. Because even if you were a knight, with the latest thing in armour worth the equivalent of $500,000 today, you could still get hurt. The musket came into favour because, though it fired slower than a longbow of similar power, any group of people could be trained to use it in a few weeks rather than the years of training required to use a longbow effectively.

About 250 years ago the technology arrived which would allow engineers to cut a series of twisting grooves along the inside length of a barrel. These grooves, called 'rifling', gave a spin to the bullet like a gyroscope and made the weapon far more accurate. These grooves also gave the name 'rifle' to a weapon employing them. By the late 1700s and early 1800s the rifle was spreading to the military, from being a gentlemen's sporting gun, through issue to 'Special Forces'. Sounds amazing having Special Forces in Napoleon's time doesn't it? Famous amongst these were the 'Rifle Regiments' of the British Army such as the 'Light Infantry' and the 'Green Jackets'. Eventually the rifle proved so superior to muskets that it was issued as standard to all infantry soldiers. Cavalry kept pistols or carbines, short rifles, for the same reason 'tankies' and truck drivers carry short weapons today – they are easier to handle or stow on your vehicle and are not used very often.

The rifle began the last 150 years as a muzzle-loading, flintlock weapon firing three rounds per minute at a maximum effective range of 100 yards in dry weather only. It progressed through a heavy, bolt-action weapon able to fire 10–20 rounds per minute out to an accurate range of over 1,000 yards during the two world wars. By 50 years ago it had developed to a light, automatic weapon capable of firing 13 rounds per second to an effective range of 300 yards.

**The modern rifle:** By the end of World War II the generals of most countries had come to the conclusion that it was no longer necessary to be able to put a bullet through both sides of a helmet at 1,000 yards because the long distance sniping of trench warfare was over. They figured that by reducing the accuracy and hitting power of the issue rifle it could be made shorter, lighter and to fire lighter ammunition with an automatic fire capability. The idea was that this would make each infantryman a source of greater firepower. See the advantages of this later when we look at localized superiority of firepower.

Generally, it takes a heavier weapon to fire a more powerful round and a longer weapon to be more accurate. Designing a less accurate and less powerful rifle, which was shorter and lighter with a higher rate of fire, was expected to make the infantry soldier more effective in battle because they could carry and fire more rounds. This new package became known as the 'Assault Rifle'.

The United States was already using the M1 Carbine by the end of World War II and this was a step in the direction of an assault rifle, being short, light and semi-automatic. Just after the war the Russians brought in the 'Assault Kalashnikov 1947' or AK47 assault rifle. This was followed by the M16 or 'Armalite' assault rifle brought in by the US Army which had all the required features.

The British were slow to change to an assault weapon. In 1957 they brought in the Self Loading Rifle, or SLR, a semi-automatic version of the Belgian FN FAL (*Fusil Automatique Lèger* – Light Automatic Rifle) that was half way between the old style 'long rifle' and an assault rifle. It was still long and heavy and it still fired heavy ammunition but it was semi-automatic rather than bolt action. As the saying goes, the generals are always planning to fight the last war.

I used an SLR, or FN – the automatic version – through much of my career and remain rather prejudiced in its favour because the FN fires the 7.62mm x 51mm 'Long' cartridge and

A selection of FN FAL (Fusil Automatique Lèger) Light automatic rifles produced by the Belgian armaments manufacturer Fabrique Nationale de Herstal. This was half way to the modern assault rifle and one which I used with great success throughout my career. (Cody)

when you hit someone with a round from an FN they go down and stay down. The effective range of an SLR was 600 yards but it weighed nearly double the weight of most assault rifles and the ammunition was almost exactly double the weight of the NATO standard 5.56mm x 45mm or Russian 7.62mm x 39mm ammunition.

For all its faults in terms of weight and length, when you hit someone with a 7.62mm FN round travelling at 838 metres (916 yards) per second they stay hit because of the immense kinetic energy in a large calibre high-velocity bullet. They go down like a tin plate at the fair ground every time, even if you only wing them, due to the massive blow of the round. This is a good thing when, for example, you meet opposition in a surprise encounter as you don't want them shooting at you after you hit them and before they go down. But both the rifle and the ammunition are a pain to carry.

In 1980 the British brought in the awful SA80 which fires the same 5.56mm x 45mm ammunition as the M16. To begin with there were terrible problems with stoppages and the rifle was loathed by the soldiers. To some extent the problems have been overcome but the rifle is still an inadequate weapon which cannot be fired left-handed and in my opinion can only have been bought by the British Army as a result of corruption. When there are so many good rifles out there such as the AK47, the M16, the Heckler-Koch G36, Dimarco and so on, why would you buy a really bad one?

## Operation: How does it work?

Almost all assault rifles are reloaded automatically, after each round is fired, by the gas from the last round. Gas-operated reloading is where a hole in the barrel near the muzzle bleeds off some of the gas propelling the bullet and uses it to drive a piston back towards the breech which in turn operates the shell-extraction and reloading cycle. Each time a round is fired, the gas drives the piston to force back the breech and working parts against a spring to extract the empty case and then allows them to come forward under spring pressure to load a fresh round into the chamber.

Almost all modern automatic rifles function on the gas-operated reloading principle because the blow-back principle used with pistols and sub-machine guns does not work well with high power cartridges and the recoil principle used with heavy machine guns such as the Browning results in a relatively slow cyclic rate of fire. The cyclic rate of fire is how fast the mechanism fires rounds when set to automatic and the trigger is held down.

So far as I am aware, all modern assault rifles have a fire selector switch which allows the user to choose between semi and full automatic fire. Semi automatic is where a round is fired each time the trigger is pressed and it must be released and pressed again to fire a second round. Full automatic is where the rounds continue to fire repeatedly at a cyclic rate of around 650 rounds per minute until the magazine is empty or the trigger is released.

## Skill: How do you use the weapon to maximum efficiency?

**Aiming and firing:** You will be taught to shoot better in your advanced training. Basic shooting is simply a matter of lining up the front and back sites with the target then squeezing the trigger smoothly while holding the butt of the rifle firmly into your shoulder. A book is not the place for advanced shooting skills but I can give you some useful pointers.

**Carrying your rifle:** When on any type of patrol or guard duty your rifle must be available for instant use. There is no point in carrying it on a sling over your shoulder. However easy and comfortable that may seem, the delay in bringing the rifle into use can be deadly. You must remember that warfare is 99% boredom and 1% adrenaline. The time you are hit WILL be the time you have slung your rifle over your shoulder and your mind is elsewhere. Then you are dead. Anyone caught resting with their rifle out of reach should be either beaten or fined a payment towards the beer fund.

On patrol your rifle should be held so that your trigger hand is near the trigger and your other hand on the fore-stock. Some units teach soldiers to carry the rifle with the butt at the shoulder but pointing down at all times. This is good for quick shooting but tiring to maintain and a soldier will stop doing this if he is not watched. I think lowering the stock to the hip is acceptable as it is the work of a moment to fire from the hip or bring the rifle to the shoulder. If you are in such a hurry then accuracy is probably a secondary concern. I don't want to teach any idle or bad habits here but if leaders make SOPs (Standard Operating Procedures) too much of a pain the soldiers will stop following them. Just carry your rifle so you can get a few rounds off quickly in an emergency. Otherwise, eventually, you will die.

**Conservation of ammunition:** When you go out on a patrol for several days on foot you will have to carry all your own rifle ammunition, probably ammunition for the machine gun and mortar bombs, grenades, explosives, flares, water, food, medical supplies, sleeping kit and spare socks. You may have to carry the machine gun itself or a rocket launcher or a piece of the mortar or a radio and possibly a kitchen sink too. This is when the weight of your ammunition counts for a great deal.

The infantry soldier is more like a pack mule than anything else. Special Forces are expected to carry more weight, further. The fitter you are the further you can walk carrying all this kit and still keep your mind clear and active. If your attention wanders because you are tired you may be dead before you rest. The loss of attention when tired is probably the best argument for fitness in the infantry soldier. Even if you are on a local patrol for a few hours or a day you still have to carry ammunition and water at the very least. On a longer patrol you are likely

Roy Mobsby, left, and Alan Ash (both ex-Parachute Regiment): bodyguards in Iraq carrying M4s (a variant of the M16), Glock 17 9mm pistols and plenty of ammunition. (Photo courtesy Sergeant Roy Mobsby)

to need more rifle ammunition to see you through a number of contacts. Especially if you are cut off from re-supply. From the above you will understand that the amount of ammunition you can carry is limited by the total weight you can carry and still function. To put it simply – the way to carry more ammunition is to carry lighter ammunition. Hence the assault rifle.

On a short patrol you might be carrying 300 rounds. At 5lb weight this is not much to walk down the street with but wait until you are running around all day in the heat while wearing body armour. On a longer patrol you might be loaded up like a mule and be carrying 1,000 rounds or more. On patrols of 5–7 days I have always tried to carry mainly ammunition and water. You don't need a lot of food in hot weather and a pot noodle weighs nothing dry anyway.

Running out of ammunition on the two-way range is like running out of air. You don't miss it until it happens. Imagine facing a mob of rioters or charging, drug-crazed fanatics or whatever is your own personal nightmare. And your rifle clicks as the firing pin comes home into an empty chamber. You feel in your pouch or jacket and there is no more. Though bayonets have many positive qualities they perform poorly against concentrated automatic fire.

Mobs are the worst in my opinion. Apparently ordinary people go crazy and will literally tear you apart. I once saw a young British soldier, working in plain clothes, dragged out of his unmarked car by a mob of Irish rioters, blinded with screwdrivers then beaten to death with iron bars – after his pistol jammed. Once a teenage soldier was castrated by a gang of Irish women. And these were ordinary decent people to all intents and purposes. They had families, children they loved and they went to church but the mob mentality turned them into monsters. Mobs still make me nervous and I've been in a good few riots so it's not just the novelty factor.

My point is that a rifle can use up ammunition at great speed if you hold that trigger down. The cyclic rate of fire on most rifles is around 650 rounds per minute so do the maths and forget the movies. When you are being shot at it is natural to want to shoot back, but a magazine of 20 rounds is gone in less than 2 seconds if you hold that trigger down. Aside from spraying bullets in the direction of the enemy when you are ambushed, try to stick to single, aimed shots.

The first time you are in a 'contact', or shooting match, you feel under more pressure as it is an unusual experience. Breathe deeply and stay calm. Fire off aimed shots and count your rounds if you have the self-control. The first time you make a kill is an even more significant experience. By my observation, soldiers all want to be by themselves for a day or so while they come to terms with the first one. After that it seems no more than turning off a light and you will be able to think a bit more clearly in the middle of a shooting match.

So, to fight a battle you have to strike a balance between defending yourself by suppressing enemy fire, which is where most of your ammunition will go, and keeping some for later to actually shoot the enemy. It is much less often that you will get an aimed shot at an enemy. Most kills will be obtained by all your team firing into the bushes that are firing at you – at night.

## REMEMBER:

Use all the ammunition you need to get out of an ambush. Use what you have to use when keeping their heads down. Use aimed shots or bursts of 2 or 3 to make single kills and always, always keep some in reserve. As your ammunition gets lower then get more careful. Running out of ammunition can be much, much worse than getting killed; it can mean you get captured...

## Types of rifle

There are a number of common rifle types available to the soldier today but basically they can be split into two types defined by what ammunition they fire. This is because a rifle, indeed any firearm, is built around the cartridge it fires. Clearly, it helps if everyone on your side fires the same ammunition so you can resupply each other.

**Rifle ammunition:** The range, accuracy and hitting power you require of your weapon determines the type of bullet you will fire and how fast it must travel. The bullet calibre and weight and the velocity required determine the design and weight of the round which propels that bullet. The size and power of the entire round determines the design and weight of the rifle which fires that round.

There are only two features which require consideration in the selection of ammunition because ammunition reliability (as opposed to weapon reliability) is now more or less 100%. The first is its accuracy/hitting power and the second is its weight. Generally, to make ammunition more accurate and powerful you have to make it heavier – by using a heavier bullet and more propellant – but to make ammunition easier to carry you want it as light as possible. The heavier the round, the heavier the weapon has to be which fires it so as to withstand the pressure and recoil. This is why snipers always use large calibre weapons – they don't use much ammo and generally don't have to carry their rifle far. So ammunition design is always a trade-off between weight and hitting power.

You could argue that the weight of ammunition is of far greater importance than the stopping power or range because you spend a lot more time carrying ammunition than you do firing it and any round making a hole in someone will stop them playing at soldiers. Plus you rarely get to shoot at anyone more than 100m away. You just have to be able to hit them. And to hit the enemy, or suppress their fire so they don't hit you, you want to carry as much ammunition as possible. The lighter the ammo the more you can carry.

A rifle bullet for use in warfare is always made up of a heavy lead-alloy core to give weight and carrying power encased in a brass-type alloy jacket to hold the bullet together inside the rifle and give better penetration of a target. This is because a lead bullet fired with the extreme power of a rifle round tends to break apart as it travels down a barrel under great pressure and flatten and spread out when it hits even a soft target.

Let me explain here that the damage done by a high velocity bullet when it hits a human body is not just the hole which it makes – though a large hole in your enemy is no bad thing. When a rifle bullet hits a target it is travelling so fast, more than twice as fast as a pistol bullet (and therefore for college educated readers the kinetic energy is squared) that it sends a shock wave out in front of it in a cone. This tears tissue and will often rip out the back of a body causing massive blood loss and certainly a rapid loss of interest in fighting you.

Pistol bullet designers aim for a bigger bullet to make a bigger hole to let the blood out quickly because, in order to be fired from a pistol, these bullets cannot travel fast enough to make much of a shock wave. As a general rule, the heavier a bullet and the faster it travels the greater the shock wave when it hits the target, the better the penetration and the better the range. But it takes a heavier rifle to fire a heavier bullet at greater velocity. What about a lighter bullet at high velocity? A lighter bullet slows down more quickly and is more likely to be deflected by wind in flight. So, as with most things in life, the choice of weight, for range, and muzzle velocity, for hitting power, against lightness of cartridge and rifle, is always a trade-off in the design of a bullet, the round which propels it and the rifle which fires it.

## TOP TIP!

### Dealing with a stoppage

The first stoppage action for most rifles is cock, hook and look. In a firefight, cock the weapon turned on its side with the ejection opening pointing down. Then as the working parts come back the stuck round will almost always drop out and the rifle will reload. If the enemy is so close – touching distance – that you have no time to reload then use the muzzle of your rifle as if you had a bayonet fitted. Push your rifle forward so the muzzle hits the enemy in the face. It will rip open their face and put them off their stroke while you think of something else to pass the time.

For a light weapon you have to either go for a light bullet travelling very fast, like the 5.56mm x 45mm round favoured by NATO and fired by the M16, or a heavy bullet travelling relatively slowly, like the 7.62mm x 39mm favoured by Russia and China and fired by the AK47. I feel that the best way of looking at the idea of firing a round which is easier to carry but does less damage is that when you have 'winged' someone they are out of the game generally and you can always finish them off at your leisure with a head shot or bayonet. That is enough. The M16/Armalite 5.56mm bullet is travelling at an extremely high velocity when it leaves the muzzle and therefore creates a terrific shock wave when it hits a human body at relatively short range. The AK47 bullet produces a lesser shock wave but a bigger hole. But in all honesty there is little to choose between them. They both do what you need.

**Ammunition in use today:** Just to make things more interesting, all bullets react in a wildly different manner when they hit someone depending on the range, if they are tumbling because they hit a leaf first, the angle they strike at, if they went through heavy clothing first and if they strike a bone in the body. What I mean is, one time a bullet will go clean through an enemy with a tiny entry and exit wound, another time a similar bullet will blow a big hole out his back, a third time the bullet will somehow bounce about inside. The bouncing around has to be seen to be believed. A little story will perhaps illustrate the 'stopping power' debate about ammunition. We were fighting line abreast through an enemy base camp in a very pleasant East African country. The camp consisted principally of trenches and bunkers. I was carrying a Belgian FN like most of the troops on our side. An American major to my left was carrying his M16 as this was what he was used to. Many soldiers, quite reasonably, derive confidence from carrying their favourite weapon.

A friend of mine, the experienced soldier and now author Yves Debay, was carrying a semi-automatic shotgun loaded with a round which fired two lumps of lead joined by a length of wire. This didn't give him much of a rate of fire but when he hit someone this weapon blew them apart. He said he liked them to go down when he hit them. His position might sound crazy if you have not seen much combat up close, but it is a fact that lead scouts in some theatres carried an RPG7 on their shoulder facing forwards. This had the apparent intention, and certainly the effect, of giving them one quick shot at you with a powerful weapon when they were hit but before they went down. You don't want to know what an RPG7 round does when it hits a human body. Actually, I suppose you might. See p.62 at the end of this section.

Back to the base-camp. We were fighting through, knocking down anything that put its head up and throwing fragmentation and phosphorous grenades into bunkers, when the major shot someone in a trench with one round at a range of 8 or 10ft. The target wasn't very happy and did not offer any further resistance, slumping down into the trench. There was an Intelligence captain close by and he asked me to take a look at the target and see if I could keep it alive. Dragged out of the trench the target was semi-conscious and bleeding a little from an entry wound high in the front of his chest and a neat exit wound in his back a little lower down.

The 5.56mm calibre bullet from the M16 had made a little hole at the front, minced one of the man's lungs with its shock wave, and then made a neat little hole as it left through his back. He was certainly out of action for the foreseeable future at the cost of half the weight of a 7.62mm FN round. A 7.62 round would probably have gone straight through (only 'probably' because as stated above bullets often do funny things when they hit targets) and come out with a lump of meat at the back but the result would have been no better.

The main problem from the target's point of view was difficulty breathing. When the chest cavity is punctured the lungs tend to collapse which prevents breathing and the casualty gurgles a lot. This is because we open our lungs through the suction created by the downward movement of our diaphragm. Air is normally sucked into the chest cavity down the throat as the lungs expand. When there is a hole in the chest the air comes in though the hole as the diaphragm moves down but instead of filling the lung it fills the air cavity around the lung and allows the lung to collapse. This is why such a wound is called a 'Sucking Wound'.

## TOP TIP!

### Dealing with a sucking wound

It is simple to treat a sucking wound and recovery figures are good – all you do is smear blood on two pieces of plastic bag and use them to seal the wounds – front and back in this case though there could only be one in some cases. We will look at this later. I did this and tied the plastic on with field dressing bandage. I then gave the man an intravenous drip against blood loss and shock then morphine against the pain and to keep him quiet. The captain tried several times to get the man out in a chopper but they were being shot to pieces by anti-aircraft machine guns. In the end someone shot the man properly to put him out of his misery.

There are just two main types of rifle ammunition in use around the world today: the 7.62mm x 39mm 'Short' round is fired by the AK47 and is in use from Russia and China to the Middle East. (There is an even shorter 7.62 x 25 round fired by a Russian Tokarev pistol and some SMGs.) The 39mm bullet has a relatively large diameter, or calibre, but a relatively low muzzle velocity owing to the lack of propellant. It is still far quicker than a pistol however, so it works by making a large hole in the target and by something of a shock wave to stop the enemy. The effective accurate range of an AK47 is around 300 yards and it weighs about 8lb.

The 5.56mm x 45mm round is fired by the US M16, the British SA80, the Italian Berretta, the Heckler Koch G36 and several other weapons issued to Western forces. The bullet has a relatively small diameter and a very high muzzle velocity through plenty of propellant, so it only makes a small hole in you and the shock wave rips your innards to pieces. The effective accurate range of an M16 is around 300 yards as the light bullet slows quickly. The M16 weighs about 8lb and 1,000 rounds weigh the same as the 7.62 x 39.

### REMEMBER:

There is actually not a lot to choose between modern assault rifles so don't lose sleep over whether your SA80 is not as good as the M16 or AK47. It's not the gun that will win battles; it's the man firing it.

**The British SA80:** In the early 1980s the British Army adopted a 5.56mm x 45mm calibre assault rifle so as to be able to share ammunition supplies with European and American forces. This rifle was called the SA80. It is the worst rifle ever issued to the British Army and probably the worst weapon ever issued to any army. It is still roundly hated by all troops despite various modifications and improvements. While it has been in use, the weapon has been known to jam and pieces break off it. There is also no way that it can be fired left-handed.

**The Russian AK47:** The AK47 is loosely put together and rattles when you shake it – but it is probably the best rifle ever made. I can say this with some objectivity because I have used it or had it used against me ten thousand times. Originally designed by Mikhail Kalashnikov at the end of World War II, the AK group – Assault Kalashnikov – comes in a range of models. There is the standard AK47 with a wooden butt and fore-stock, there is another which has a wire folding butt and there are several more.

Kalashnikov's idea of the features to be found in an ideal rifle was exactly the same as all modern thinkers but he was an extremely clever engineer, which is clearly visible in the end result. We have covered these before but I shall go over them again to help them stick in your memory:

- The AK47 is light for ease of carrying
- The AK47 is short for ease of use in a building or vehicle
- The AK47 fires 7.62mm ammunition which is light yet stops a man quickly
- The AK47 has a magazine which holds 34 rounds and feeds them without jamming
- The AK47 is utterly reliable and will fire after being dipped in mud
- The AK47 was also designed to be easy to manufacture – which is a concern to a country's planners when going to war and speed of production is an issue

Remember the AK47 round is 7.62mm x 39mm and this is different to the longer 7.62mm x 51mm fired by the SLR or FN in that it has a bullet of similar weight and diameter but a smaller cartridge and less propellant. This does lower the muzzle velocity and reduce the range but it will hit hard enough at 300 yards while still only weighing the same per thousand rounds as 5.56mm ammunition. The relatively large bullet travelling slower still makes a large hole

An AK47 – originally designed towards the end of World War II it is still widely used throughout most Third World countries. (Sean Clee © UK Crown Copyright, 2007, MOD)

in a target and is not so easily deflected by wind, twigs and other cover en route to the target as the 5.56mm.

The AK47 is purposely made with loose working parts so as to prevent jamming caused by sand, dirt and carbon from the propellant. It was actually designed to be carried and used by peasants fighting unsupported and with insufficient discipline to ensure regular cleaning. It is a well known fact that an AK47 can be left in a pool of mud for two weeks then taken out and fired. Try that with some modern rifles.

**The American M16:** The M16 is very light – even lighter than the AK47 – and has an automatic fire capability insofar as some versions have full automatic and some allow the firing of three round bursts. It fires the 5.56mm x 45mm round which hits hard enough and is light to carry. The M16 has only one real fault: when it is dirty it jams. It jams from a build-up of carbon if it is not cleaned regularly and it jams if it gets sand in it as many men have found to their cost in the Middle East. For God's sake keep your M16 clean – and then it will look after you.

Reliability has been much improved since early problems in the Vietnam War where rifles were frequently jamming due to heavy use and the troops initially being told they did not

A US Special Forces detachment meet with General Tommy Franks in October 2001. They are armed with a M11 sniper rifle on the left and a M4A1 carbine (including sound suppressor and optical sight) on the right. The beards and traditional headgear all help as well. (DoD)

need cleaning — can you believe that? In essence they made a beautifully engineered weapon with tolerances like a Swiss watch.

The US Army are now planning to replace the old M16 with the Objective Individual Combat Weapon, or OICW. It will fire the same ammunition but be built to work in a dirty or sandy environment. To improve use in and from vehicles the length will be reduced from 40in to around 30in. In the meantime troops are being issued with the M4 which is essentially a M16 with a folding stock and a cut-down barrel. This will help improve use in vehicles but you must still keep it clean. Good move General.

## The future of the rifle

In conventional warfare the rifle has now come close to the peak of its potential insofar as it is light, reliable, fast firing and easy to maintain. It has reached the place in its development comparable to the modern motorcar, which, after 100 years of rapid development has now reached a level of sophistication, which will only see minor improvements to the current product, better ashtrays or whatever, before we would have to go back to the drawing board and start again.

## Left- and right-handers

Out of every 11 men, around 10 are 'right-handed'. Due to the benefits of standardization, and millennia of custom, the world is set up for 'right-handers'. This puts left-handers at an apparent disadvantage in many fields. Obviously, for design reasons, it is necessary to have the cocking handle on one side of a rifle and so forth but the point is that left-handers can get around this – they just need the weapon to be fire-able with the left hand. The SLR, FN, Armalite and AK47 can all be fired by left-handers from the left shoulder. The SA80 cannot. You may not care much about the happiness of left-handers but the firepower of a unit is reduced by every man not comfortable with his weapon.

Much more important than the happiness of left-handers is that when you want to fire round the corner of a building or other cover half the time you will need to fire left-handed to avoid exposing your body. It goes without saying that all soldiers must learn to shoot with either hand. So now you might see why I am so incensed that the British Army selected a weapon which physically cannot be fired from the left shoulder? Have these guys never been in a firefight? If so why are they making these decisions?

German soldiers with the Heckler & Koch G36. In my opinion this is marginally better than both the M16 and the AK47. (Topfoto)

To improve the rifle further would require a reduction in flash, noise, weight or ammunition supply and none of these are likely to be achieved so long as the rifle remains in anything like its current cartridge-fed, explosive-propelled, projectile-firing form. The next real generation of 'rifle', after the assault rifle, will probably be some form of beam or particle weapon. This could feasibly be silent and flash-free, thus overcoming the two major remaining drawbacks of the modern rifle. I have heard from a reliable source that the US is working on this.

## Summary: What are the important things to remember about rifles?

I've talked on and on about rifles and their points for and against. The fact is you probably won't get much of a choice in your rifle as long as you are a junior rank. What I want to get across to you is that all the rifles you are likely to see or use are pretty much all OK. They all weigh next to nothing and fire light ammunition very quickly, reliably and accurately. Keep your rifle clean, whatever it is, and it will serve you well. What you do need to consider is that it is best to use the same ammunition as everyone around you – in case you or they run short.

If I had to choose the Heckler & Koch G36/G36E is slightly better than either the M16 or the AK47, and probably demonstrates what the US Army is looking for to replace the Armalite, but is not on general issue.

Both the Armalite and the Kalashnikov are lovely weapons to use. But as a Special Forces operator I would suggest you favour a heavier, longer-range rifle in open country, an assault rifle for general use and possibly something even shorter for house clearing and Close Quarter Battle. But I still maintain that the most, indeed the only, important thing about a rifle is the man holding it.

# BAYONETS

## Goal: When is the right time to employ this weapon? What is it good for?

I hope you are never in a bayonet charge as it is an act of desperation in a modern conflict. The idea is to close with the enemy and, rather than shoot them from a distance, stab them to death

Bayonet practice, old school style. (IWM H18462)

with a long knife attached to the end of your rifle. Clearly this is a worrying prospect for anyone on the receiving end so one effect of a bayonet charge is to make the enemy break and run. About the only time to use a bayonet charge in modern times would be if your unit was running very low on ammunition and you were up against an enemy at very close quarters.

I have been in two bayonet charges. To paraphrase the Duke of Wellington when talking about his own troops, 'I don't know what effect they had on the enemy but they scared the life out of me.' Both charges were acts of desperation and in both the enemy got up and ran from positions where they could, and should, have cut us down before we got to them.

At one point we were patrolling from village to village somewhere in Africa. It was our job to go round to all the villages, show them we were friendly and try to get some information out of them about the movements of the terrorists who were living off them. Such 'Hearts and Minds' campaigns will sound familiar to Vietnam vets as well as soldiers in Afghanistan today.

In this countryside it was normal for a terrorist force to oblige the local population to feed the locals and to provide them with women. This meant the terrorists could operate freely without a supply line for food. If the villagers did not cooperate they were killed, horribly.

If they gave any information away to the peacekeeping forces they were killed, horribly. A situation not too dissimilar to what is often experienced today in rural Afghanistan. We, as the defenders of democracy, were bound by our rules of engagement to be nice to the villagers. We just gave them medical aid and so forth. 'Hearts and Minds' as they say. Nichollo Machiavelli said words to the effect that if you can only have people love you or fear you then to fear you is best. And as a famous US officer or president – depending on who you believe – once said, 'When you have them by the bollocks their hearts and minds will follow.'

If you were a villager would you toe the line set by the terrorists or would you help the 'good guys' and watch your family raped and tortured to death? In a war it is always the civilians who suffer the most and I could never blame the villagers for what happened next. Anyway, a stick* went into the village, two sticks stayed outside as per normal SOPs. When the talking was done it was the turn of my stick to go in and get some water. The first stick stayed in the village and the other stick stayed in cover outside.

As we left cover to walk towards the village, a distance off some 200 hundred yards across flat, bare, sloping ground, around 50 AK47s and a couple of RPD or PK machine guns opened up on us at a range of just 75 yards. The position for the ambush was perfect. We were caught out in the open with nowhere to go. Away from the fire was up hill with no cover. To either side was no cover for a hundred yards or more. Towards the enemy was no cover to the hedge-line they were hiding behind. In my own defence I must say that the village was surrounded by open space and that the approach I chose was the best of a bad job.

The ambush was perfectly constructed and the timing was exact. Had the opposition been able to shoot accurately we would have been dead in a couple of seconds. As it was we benefited from a feature of some of the opposition's shooting technique. They lay down in position, pointed the rifle at you and then closed their eyes and looked at the ground while they were shooting. This was not typical of the trained African soldiers I had the pleasure of serving with. My friend Corporal Gosho ('rat' in Matabele) once dropped a moving target between the eyes, at over 100 yards, from a standing position in the bush. Thank God he was on our side.

The SOP for being ambushed is to charge the ambushers – as a properly constructed ambush should be inescapable and you should die quickly so, it could be said, this tactic gives you something to do while you are waiting. Amongst my team were a couple of Cockneys who

* A stick is a small unit or section of soldiers, usually around 8 men.

had just left the French Foreign Legion Parachute Regiment and my good friend Yves Debay, who has no obvious concept of fear, so I could rely on them. Without thinking I gave the order to charge and we ran, line abreast at the enemy.

Commanding the other stick on our flank Corporal Ndlovu ('elephant' in Matabele) thought quickly enough to give us covering fire to help get the ambusher's heads down. It was certainly his action that saved us. As it was, the sight of a bunch of crazy men running at you with bayonets fixed has a certain effect. Considerably more, I think, than rifles alone. The ambushers got up and ran for their lives. If I was a cat that would have been another life used up.

## History: How did it develop to be the way it is?

Before the infantry had muskets they had spears (technically known as halberds and pikes) and the appropriate drills to use these effectively as a group. When muskets were first issued they had a very slow rate of fire of perhaps three rounds a minute on a good, dry day with well trained and steady men. If the enemy were approaching you quickly then there would come a time when they would stop you reloading. Likewise, if you were approaching the enemy and discharged your weapon you were left simply with a club.

Once this had caught on it seemed a good idea to make the spear as long as possible so bayonets became fairly lengthy. There was even a type called the sword bayonet which could be used like a sword when detached. As it is much preferable to stab someone from a distance where they cannot reach you a sort of arms race developed where bayonets became longer and longer. The only limit was the convenience of carrying and fitting them.

The bayonet is still included on all modern assault rifles. When US and British Special Forces fought in the battle for Tora Bora in 2001 bayonets came in handy. Their effectiveness was again demonstrated when Argyll and Sutherland Highlanders bayonet charged a position held by the Mahdi Army in Iraq 2004. (USMC)

The idea of a bayonet being rammed into your belly is pretty unpleasant and one of the main benefits of the weapon. To improve the fear factor some armies developed bayonets with serrated edges likes saws which look truly awful. Some retained knife-like weapons as these could be used for other purposes – like opening cans of beans – while yet others made the bayonets round or triangular in section for strength and ease of manufacture.

## Operation: How does it work?

The idea of a bayonet is that a knife on the end of a rifle turns it into a spear – which is better than a club because you can kill someone from a little further away. If you push a bayonet into someone it will quickly take the fight out of them, which is a good thing. Early bayonets were 'plug' bayonets which means they fitted into the end of the barrel and stopped the rifle being used again. This has obvious disadvantages so bayonet fitting was developed where a bayonet is clipped parallel to the end of the muzzle and remains there while the weapon is fired.

Something a lot of people don't know is that when a modern rifle is fired the barrel flexes. It whips as a shock wave travels along it. Only a tiny amount but enough to affect the flight of the bullet as it leaves the muzzle. The length of all rifle barrels is tuned so that the barrel is in a suitable stage of its wave cycle as the bullet leaves it. When a bayonet is attached to a rifle it affects this wave action and can result in a significant effect on the direction of the bullet. I was told as a young man that a bayonet fitted to a World War II Lee Enfield .303 made

## TOP TIP!

### Using your bayonet effectively

There is one little trick which might come in useful. You will understand that 'reach' is very important in being able to stab someone before they stab you? If you hold your rifle at the 'on guard' position with a sling fitted then it is possible to move your rear handgrip from the narrowing of the stock to where the sling attaches at the rear of the stock. You can then, at the right moment, release your front hand and push the rifle out in front of you with the rear hand. This extends the reach of your bayonet by around 3 feet and could make all the difference in a tricky situation.

it shoot 10in low at 100 yards. On a modern rifle, even if the effect is not so pronounced on the barrel flex, the weight of the bayonet will certainly affect your aiming. My point is that you only fix bayonets when you are about to use them or you are so close to the enemy that accuracy is not an issue.

## Skill: How do you use the weapon to maximum efficiency?

Close with the enemy at the run while screaming. I think that ought to just about do it. The idea is that this has the effect of making the enemy run or freeze in fear. In the old days, just dying out as I was trained, soldiers were taught bayonet fencing which, as the name suggests, is where two men face off with bayonets fixed and try to stab each other. With automatic weapons this is not going to happen as you would both have to be jammed or out of ammo with no pistols, grenades or mates around – an unlikely occurrence.

## Types of bayonets

Each rifle in production has its own bayonet designed for it so there are simply thousands of differing lengths and styles. The main types are the spike, which is a round or triangular section spike, the knife, which is effectively a knife with a bayonet fitting and the sword, which is a very long knife. As a general rule, modern bayonets are shorter than those in use during or before World War II. The Kalashnikov range actually has one model which has a bayonet attached all the time. You just swing it forward and it locks in position. It also forms a wire cutter – which is pretty clever.

## Summary: What are the important things to remember about bayonets?

The chances are that you will never use a bayonet in anger. But, so you don't dismiss the bayonet out of hand, let me tell you a little story to close this section. In Iraq, 2004, there were a couple of soft-skinned vehicles carrying British soldiers through Basra. The Shiite Militia enemy there were told by their leaders that the British were cowards and would not stand and fight so the Militia set up an ambush. Actually the British rules of engagement forbid shooting except in special circumstances at that stage of the conflict and British soldiers always follow orders. There were about 100 insurgents in the ambush and they believed that they would be able to shoot at least a few British soldiers and the convoy would continue driving – effectively running away. There is wrong and there is wrong.

The men in this little convoy were Scottish soldiers: the famous Argyll and Sutherland Highlanders. Twenty of them – which is more Scotsmen than you ever want to fight. At the battle of Balaklava in 1856 the Argyll and Sutherland Highlanders' colonel, Colin Campbell, had to hold them back from breaking formation to rush in and bayonet the Russian cavalry they had just fought to a standstill with the very British cry '93rd, 93rd, Damn all that eagerness!' Back to Basra: finding themselves under enemy fire the Jocks stopped their vehicles and disembarked to form a defensive perimeter while calling for reinforcements on the radio. As these were going to take a while to arrive, and the Jocks have little patience, their officer gave the order to fix bayonets and 20 Jocks charged the remaining 85 enemy across 600 feet of open ground to bayonet 20 of the enemy on contact beside the 15 they had already shot. There would have been more enemy killed had they been able to catch them. Seems the Jocks haven't changed much since Balaklava.

# HAND GRENADES

You will have seen 'atomic' hand grenades in the movies – the grenade goes off and bodies and vehicles go flying everywhere in sheets of flame. The reality, as so often the case, is a little less impressive but a grenade is probably your best weapon after your rifle so listen up. There are three main types of hand grenade you should know about and they have slightly different uses so get the info below into your head...

## Goal: When is the right time to employ this weapon? What can it do?

All hand-thrown grenades consist of a charge surrounded by a layer of something which turns into shrapnel, gas, flash or smoke when the charge goes off. They all have a pin which

Russian fragmentation grenades clipped to belt of Afghan fighter in 1992. (Corbis)

retains a lever which, when released, sets off a fuse of some 4.5 seconds to delay the detonation until you have thrown the grenade. The main types of grenade you are going to come across are these:

- The explosive fragmentation grenade which explodes and sends splinters of steel flying in all directions
- The phosphorous grenade which is officially for signalling but really for spreading a cloud of burning phosphorous and poisonous gas around
- The stun grenade which is supposed to be to stun the bad guys long enough to arrest them while leaving the hostages unharmed. Yeah, right.

There are several other types of weapon, listed below, which are called grenades but which are either obsolete, useless or for special occasions only.

In essence, fragmentation grenades are for clearing trenches, bunkers and rooms without entering them. Phosphorous grenades are for clearing bunkers where a bend or a sump might protect the occupants from a frag grenade. Stun grenades are, as stated above, for stunning the enemy.

## History: How did it develop to be the way it is?

Grenades started to be used in the 1400s and at this time were iron balls filled with gunpowder and detonated by burning fuses which stuck out – rather like the comedy bomb you might have seen on cartoons. The name comes from *granado* which is Spanish for the pomegranate which some think it resembles.

You can probably imagine that in a time of extremely slow rate-of-fire muskets and swords a thrown bomb was a pretty handy tool. By the 1800s the grenade had gone out of use but was revived in the Russo-Japanese War of 1904–5. Then there were a bunch of homemade efforts until the English inventor William Mills invented the excellent Mills Bomb in 1915, the pineapple grenade which has continued in use pretty much to this day. In World War I the British attached sticks and used a long fuse to make them easy to throw a long way. But the brave Germans threw them back so the fuse was reduced to the now standard 4.5 seconds and the stick binned. The Germans picked up the stick idea and developed it for use in World War II.

Since World War II the idea has arisen that by surrounding the charge with brittle wire marked into segments, rather than a cast-iron case, the shrapnel could be shared out more fairly with everyone getting a number of small pieces rather than one or two people getting big ones. I don't see much difference in practice and I have used many different kinds of grenade. Indeed, I caught 13 pieces from a Russian fragmentation grenade and though it was the cast-iron type they were all tiny – perhaps a quarter of an inch long. They did sting a bit though.

## Operation: How does it work?

**A fragmentation grenade** is not supposed to kill the enemy. It is unlikely to kill someone unless it explodes right on them or they are unlucky enough to catch a large piece. What fragmentation grenades are for, and what they do very effectively, is stun people long enough for you to get up close and despatch them with bullets or bayonets. Or, failing this, they make people seek a medic, a casevac and a ticket home. A fragmentation grenade is useful at night because it doesn't give your position away and for clearing trenches, rooms or bunkers. Toss it in, wait for the bang, walk in and finish off the wounded.

The average fragmentation grenade is an ounce of explosive surrounded by something which will break up into shrapnel and spread itself around violently so everyone gets a piece.

World War II grenades such as the British Mills 36 were cast-iron hollow pineapples weighing 27 ounces. Into this was packed an ounce of explosive with a hole up the centre to fit a stick fuse and detonator. Modern fragmentation grenades tend to have a similar charge surrounded by a thick layer of brittle wire. The theory is that a cast-iron body produced a few large lumps of shrapnel which might miss the target. With thousands of tiny pieces of wire everyone around is served equally. A sort of military democracy.

**Concussion grenades** are a larger charge but with no shrapnel. They are supposed to be more effective in a confined space but few soldiers are going to carry several types of grenade. So pretty useless and the sort of thing a bean counter would design.

**A phosphorous grenade** is unlawful when used against the enemy! They are outlawed by the Geneva Convention as they are 'nasty'. But they are best for clearing bunkers where a person might somehow be shielded from the blast of a fragmentation grenade. If anyone stays in a bunker with a phosphorous grenade they will die, blinded and burnt, and if that doesn't do it the poison will. A phosphorus grenade looks like a small aerosol can with the ubiquitous lever down the side. It is light and is used by taking out the pin and throwing – a long way. When it detonates it makes a gentle 'pop' and produces no shrapnel. What it does do is create an instant cloud of white phosphorous smoke filled with flying, burning lumps of phosphorous. The smoke burns lungs to a fatal degree and the lumps burn through flesh if they settle on it and set fire to wood. An immoral or amoral soldier might use a phosphorous grenade as an extremely effective and safe way of clearing a bunker. He would not have to go in but just shoot anything that blindly staggered out, which would in some ways be a kind of mercy killing!

**Smoke grenades** can be used to signal aircraft or to block the enemy's view while you advance so they do have some use. Of course, you can also signal with a phosphorous grenade... Tanks actually carry Phosphorous shells for signalling if all the five radios break down and the radio tech cannot fix them.

**Incendiary grenades** are similar to phosphorous grenades but they employ a chemical reaction to burn at a high temperature – 2,200° – which is handy for setting fire to things but a phosphorous grenade will do that too.

**Anti-tank grenades** were developed in World War I and were suicidal for obvious reasons. They are now all obsolete. The early efforts were several fragmentation grenades in a bag. Eventually the Germans came up with a clever stick-on blast weapon which I will not explain here as a derivation could have insurgency applications. You will never tackle a tank with a bomb as the modern anti-tank weapons are so much better.

**Stun grenades,** or flash-bangs as they are sometimes called, make a very loud noise and a flash. Hence the words on the tin. The stun grenade is usually only issued to Special Forces and police when doing hostage rescue so you may never see one. I certainly hope you don't get to see one as they are a waste of time. The idea is that you throw one into a room filled with both hostages and terrorists. The grenade makes such a loud bang and a dazzling flash that all the rooms occupants are disoriented long enough for the good guys to get in and shoot the bad guys. You can always tell the bad guys – they wear black hats. They do actually stun most people who are not expecting them but, in my opinion, it is often better to use a fragmentation grenade as a frag will *definitely* stun people and you can always patch the hostages up afterwards. They presumably don't have a bus to catch.

Throwing a stun grenade. Although issued to SF soldiers for hostage rescue I generally consider them a waste of time. (Barry Lloyd © UK Crown Copyright, 2010, MOD)

If you ever have to resolve a hostage situation on the fly, so to speak, toss in a standard frag grenade especially as you are unlikely to have any stun grenades.

**Tear gas grenades** are used against rioters and have the effect of making the eyes run and throat burn a little. They are only marginally effective as it would take concentrated gas in a closed room to truly incapacitate anyone. Puke gas (exactly what the name implies) is far better as even a whiff of this will have even the most hardened rioter bend double throwing his guts up rather than throwing rocks at you.

**Sting grenades** use rubber balls instead of shrapnel to cause pain to rioters and are generally a waste of time.

**A Molotov Cocktail** is usually what people call a bottle filled with petrol and a burning rag in the top. The real McCoy is a little cleverer but I had better not explain how to make one here as this information could easily be used by the wrong kind of people. It does not need lighting and bursts into flames on impact. Its obvious targets are humans, vehicles and buildings.

## Skill: How do you use the weapon to maximum efficiency?

To use any grenade the lever, which lies snugly down the side of the body of the grenade, is held in place by hand and the retaining pin is removed. Nothing will happen until the lever is released and prior to this the pin can be reinserted safely should you wish. When you throw the grenade, and thereby release the lever, the lever flies off and a striker hits the fuse which, after an interval of 4.5 seconds, sets off the charge. By this time the grenade should be well away from you.

Fragmentation grenades work most effectively, like mortars and artillery shells, when they explode in the air. With practice it is possible to release the lever, let the timer start and wait for a certain interval before throwing – with the effect that the grenade explodes while still at head height amongst the enemy. There is a knack to this of course.

So if you want to clear a trench without getting too close, and maybe shot, then throw a grenade in first and follow it before they have time to come round. Like shooting fish in a barrel. Likewise with houses and bunkers – though with bunkers you might want just to keep throwing grenades in as following them is not always a good idea – you don't know what you might find in terms of booby traps and other 'nasties'.

There is one other excellent feature of grenades which I want to share with you here. Most fighting happens at night and not everyone has night sights. Hopefully at least the enemy don't. What tends to happen without night sights is that two sides face each other in cover. When the first person fires on one side then they light themselves up with the muzzle flash from their rifle and everyone on the other side fires at them for a while. Then the firing dies down and after a while the process is repeated. Of course no one wants to fire first. This is where grenades come in. If you can work out where the opposition are you can lob grenades at them all night without giving your position away.

One time we were in the Zambezi valley, 18 of us dug in like rats in a little triangle, 10 yards on a side, with a machine gun at each corner. There was moderate bush all around. Each night at 18:30 it went dark and we were attacked by about 200 of the opposition. It felt a bit like Rorke's Drift – if you remember the film *Zulu*. Each morning we did a clearing patrol and from the piles of empty cartridge cases we could count how many were shooting at us – they tended to lie still and shoot off all their ammunition leaving a neat pile of cases – and from the blood stains we could see how many we had actually hit.

Fortunately it is one of those strange quirks of human nature that a person lying on the floor shooting at night, and without night sights, tends to shoot high so we had no casualties. And we knew to shoot low to compensate. This went on for four nights with someone on one side firing first and everyone else shooting at them and so forth. Each night we hit about 20 of the opposition. It so happened there was a crazy American lieutenant amongst us. One night he suggested he and I crawl out towards the opposition so we could nail them properly instead of aiming at muzzle flashes. Even today what follows sounds crazy to me. Anyway that's exactly what we did. We crawled up towards where the opposition were shooting at our position from – some 25 yards from our trenches – and got within about 5 yards of them. They were just on the other side of a raised dirt track and shooting over our heads. I tossed the first grenade over and judged a hit by the gurgling. A second grenade and the gurgling stopped. Then a whole bunch of enemy walked up the track in the dark and stumbled over us. We lay on our backs with me throwing grenades in all directions and the lieutenant shooting everyone who tripped over him and no one knowing what was happening. Remember it was pitch dark and every shot and grenade flash lit up the area like a freeze-frame film for a split second and then left everyone dazzled.

But there were bodies everywhere so in the confusion we were able to crawl back to our trenches. The point to remember here, the real moral of this story, is that grenades are anonymous at night and rifles are certainly not. No one knows where you are throwing grenades from at night even when you are lying at the feet of the enemy.

## Types of Grenades

There are a thousand types of fragmentation grenade but they all either look like a pineapple or a lemon and they all work the same way. You pull out the pin while you hold the lever in place. When you release the lever you have around 4.5 seconds to part company. The pineapple is a cast-iron shell which fragments on detonation and the lemon is a tin casing which covers a segmented wire coil which does the same trick but more delicately.

**LEFT TO RIGHT** Modern US fragmentation grenade; modern German frag; older Russian frag; older US frag. (i-Stock)

## Summary: What are the important things to remember about grenades?

Carry plenty of fragmentation grenades and perhaps a couple of phosphorous as they reduce your risk when clearing trenches or buildings and when fighting close up at night. Grenades provide the only safe way to enter a room no matter what you have seen on the TV. If in doubt, toss in a grenade first. If you think there may be civilians inside you may shout a warning first telling them to come out. They may elect to blow themselves, and you, up at this stage.

# MACHINE GUNS

Machine guns come in many models and calibres but they only have two functions: the first is to put a lot of bullets into one area quickly and the other is to stop light armour or knock down walls. The definition of a machine gun is, I always think, something along the lines of a firearm, which can fire rifle or heavier ammunition and reload and fire again repeatedly on an automatic cycle for as long as the trigger is depressed. In practice they are all fed by a belt of ammunition rather than a magazine.

This term, though, is never used for automatic rifles as they have magazines and therefore limited duration or sub machine guns/machine pistols for the same reason plus they fire pistol bullets. A bit vague but you will know one when you see one. A machine gun is generally anything from a heavy barrelled automatic rifle of 7.62mm calibre to a light cannon of perhaps 27mm which fires explosive rounds, but above 12.7mm weapons are often called cannon or grenade launchers, particularly if they fire explosive rounds. To some extent the term is blurred and it depends who is talking. The thing which really defines a machine gun is that with one pull of the trigger it produces a high rate of fire and can keep firing for an extended period – usually some thousands of rounds. Machine guns of various light calibres are carried by infantry while heavier weapons are mounted on vehicles, aircraft and boats.

## Goal: When is the right time to employ this weapon? What can it do?

The problem with a machine gun is that it is heavy. It has to be to fire the large numbers of heavy rounds, and dissipate the heat thus generated, and it has to be accompanied by a lot of heavy ammunition. For this reason it is best fitted to a vehicle or sited in a defensive position with a good view of the approach. The machine gun is ideal for shooting down a wave of charging infantry – which is why infantry don't charge much anymore.

Since World War II a light or medium machine gun with a calibre in the range of 7.62mm has usually been a part of every infantry section of eight men or so. This added a great deal of fire power to a bunch of bolt-action rifles by raising the combined rate of fire from the rifles, perhaps 80 rounds per minute, to a thousand rounds a minute. It was ideal for stopping an enemy charge or keeping their heads down while you charged. In recent times, as infantry have been issued with automatic rifles having a higher rate of fire, the machine gun has become less of a clear advantage to a section of infantry when the fact of its weight has been taken into account.

I would recommend that a machine gun no longer be carried in an infantry section but that 7.62mm machine guns should be available on vehicles and supplied to static positions for use in their defence. Probably the advantage of a machine gun on a vehicle is that it is impossible to shoot accurately from a moving vehicle and so a machine gun gives you a better chance of hitting something. It also has a higher firing position to lay down suppressing fire but this is a two-edged sword in that it also makes a good target. In a defensive position, there is often little concern for weight of ammunition so the ability to lay down suppressing fire on the attacker becomes worthwhile.

**The machine gun on patrol:** When a light machine gun is taken out on patrol it has two folding legs under the barrel or fore-stock, which support the weapon at the best height for use when the gunner is laid behind it. These legs can be gripped as one handle for when the gun is fired from the hip in close combat. Wearing the sling across the opposite shoulder at the same time gives greater stability.

Historically two men have generally been assigned to the gun in an eight-man section – the gunner, to carry and fire it, and the commander, to carry extra ammunition and spot

Parachute Regiment on patrol, Afghanistan 2010. A machine gun team take up a fire position. (Photo courtesy Tom Blakey)

A foot patrol of Royal Marines from Delta Company, 40 Commando passes near Kajaki, Afghanistan. (Si Ethell © UK Crown Copyright, 2010, MOD)

targets or defend the gunner. The commander, often a corporal, also carries a rifle and at least 1,000 belted rounds on a short patrol.

When a patrol is 'surprised' into action, by ambush or sudden confrontation, the members take cover separately and the two men on the gun often get parted which stops the commander getting the reserve ammunition to the gun and rendering it less useful. The British issued GPMG (General Purpose Machine Gun) weighs 27lb and the US M240 a little less, (both excellent 7.62mm medium machine guns) nevertheless, in the past I have thought it a good idea to put a strong man on the gun, let him carry no other stores and rest him regularly, but have him carry 500 or 1,000 rounds as well. Of course, this requires a fit and strong gunner but many gunners take pride in their ability to carry their gun.

**The machine gun in a 'fixed fire' role:** In a fixed position you don't have to carry the gun so you can fit a heavier barrel – to stand the heat in case you have to fire at waves of attackers for long periods – and you can do something else which is quite clever: you can set up the gun on a special tripod which gives the facility to record the direction and elevation of the gun as it is test-fired at various likely targets in daylight. These might be ditches approaching the position or gaps in minefields or wherever the enemy might gather or assault in strength. These target settings are then kept safe and ready for when you are attacked during darkness, fog or smoke. Even if you are only guessing where the enemy are in their approach, or maybe you can hear but not see them, a good burst onto one of the recorded targets can really spoil the enemy's evening.

The 12.7mm/.50in heavy machine gun is effective against buildings and can easily pass through two walls at 2,000m. Remember this if one is being fired at you. It is also effective against light armoured vehicles, particularly if supplied with armour piercing rounds. In counter-insurgency warfare you will not come across much enemy armour but if you have the vehicles try to keep a heavy machine gun handy for knocking down walls and buildings sheltering insurgents. I once carried the barrel of a 12.7mm (.50cal) Browning over some mountains for four days and, trust me, you do not want to carry one as infantry. Of course insurgents in Afghanistan often have Russian 12.7mm guns so take this into account with your armour and defences.

## History: How did it develop to be the way it is?

Before the invention of the machine gun, infantry fired individual shots at each other at a rate of three rounds per minute per man if they had muzzle-loaders and slightly faster if they had an early breech loader. By the time of the American Civil War the all-metal cartridge, required both for a breech-loading rifle and for a machine gun, had just been invented but was not widely available. Prior to this a paper cartridge was bitten open, the bullet at the top kept in the mouth as the powder was poured down the barrel. Then the bullet was spat down the barrel followed by the paper cartridge as padding.

During the American Civil War the first machine gun appeared, having been invented by a Mr Gatling in 1861. This was hand cranked and fed by a long vertical magazine above several rotating barrels. Despite the disadvantages of generating a huge cloud of black powder smoke, and being a target for artillery they could not reach, the advantage of a weapon able to fire a couple of hundred rounds a minute against three rounds can easily be imagined. It made the gun equal in firepower to a company of men. Gatling guns were soon used to win empires all over the world.

A few years later the first self-powered machine gun with a single barrel was developed by Mr Maxim and this was quickly copied by Mr Vickers and others. By 1900 all machine guns were single barrel as this was lighter and cheaper to manufacture. By the outbreak of World War I in 1914 the machine gun had become the biggest killer of infantry in the open. Before World War II Mr Browning had devised the machine gun which bears his name and which is still in use today pretty much unchanged in .30in and .50in (12.7mm) forms.

## Operation: How does it work?

There are three issues to be resolved in designing a machine gun: **Automating Action** (How do you make the gun complete the load, fire, extract cycle), **Ammunition Feed** (How do you bring the ammunition to the breech area) and **Cooling** (How do you stop the barrel melting – yes, the barrel really does get that hot). I have seen barrels glowing cherry red in the night when fending off a sustained attack. At times like this a gunner may be less concerned with avoiding melting the lining of his barrel and more concerned with health and safety. His own health and safety that is.

**Types of automating action:** There are three main types of automation used to power the firing cycle of a modern machine gun. These are gas, recoil and electric.

Many lighter types of machine gun – such as the 7.62mm – use what is called the 'Gas System' similar to that which operates assault rifles. In principle this involves a gas port near the muzzle of the barrel, which bleeds off a certain amount of gas from the firing of each round after the bullet has passed on its way to the muzzle. This gas is then channelled to push back a piston, against a spring, which pushes back the 'working parts' to extract the empty cartridge case and eject it. The spring then forces the working parts forward to pick up a new round and feed it into the breech. The gas system is fast and efficient but the working parts do get dirty quickly from the propellant. Black, hard 'carbon' forms on many of the

Santa behind a Browning .50cal in Afghanistan, 2010. Spare barrel to right. (Photo courtesy Tom Blakey)

working parts, particularly the piston. This must be cleaned off regularly or the gun will jam. The gas power, which is available to drive the cycle, can be adjusted by the user so as to raise the rate of fire or compensate for a dirty gun running slow or sticking.

The old style .30in and .50in Browning machine guns, together with some heavier machine guns today, use a form of recoil action where the energy from the recoil of the barrel is transmitted to the working parts which causes them to move backwards, extract the empty case, pick up a fresh round and feed it into the breech. These tend to fire slower than gas operated guns but they are very reliable.

Some of the latest high-speed machine guns are powered externally by electric motors so as to achieve very high rates of fire. A very high rate of fire becomes useful when you have only a short time to fire such as a ground attack aircraft passing over the target.

**Ammunition feed:** To sustain a high rate of fire it is impractical to feed from a magazine as it cannot contain enough rounds to last for more than a second or two. The answer is to feed from a belt of some kind. The belts of ammunition, which feed all modern machine guns, are formed from what is called a 'Disintegrating Link'. This means that the rounds are joined together in a flexible chain, each link being a round fastened by a metal clip to the next. As the mechanism of the gun takes the round to load it into the breech, the round, which acted as a pin holding two pieces of chain together, is withdrawn causing the chain to part into sections. By this action the output of the gun is a stream of bullets out of the muzzle, a stream of metal clips out of the side opposite the ammunition feed and a stream of empty cases out of the bottom. Ammunition for machine guns is normally supplied in metal boxes containing one belt of 400 or 1,000 rounds.

**Cooling:** This has always been a problem with machine guns owing to the heat generated by the firing of the propellant and the friction of the bullet along the barrel. There have been a whole range of systems employed to combat this problem with varying success. Some infantry guns have a light barrel for patrol use and a heavier barrel for the sustained fire role. Barrels can be changed in seconds with a clip and a twist so two or more barrels can be rotated in a sustained fire role.

Some barrels are made with fins to dissipate the heat better and some have linings of chrome or other heat resistant steels. Some high-speed guns even use five or more barrels firing in turn like the old Gatling gun which allows the barrels time to cool between shots. Probably the most effective way to cool a gun is a water jacket but this is prohibitively heavy for carrying. The water is pumped around the barrel in a cycle which cools the gun barrel and then cools the water away from the gun.

**Tracer rounds:** One in five or so of the rounds in the belt fed to a gun is what is called a 'tracer round'. In the back of the bullet is a chemical which lights up and makes the path of the bullet visible from behind in daylight and more so at night. Tracers are commonly designed to light up only after travelling 200 yards so as not to give away the position of the gun more than necessary – at least in daylight where a clean, dry gun cannot be seen by smoke or muzzle flash – and to burn out after about 900 to 1,100 yards depending on the round.

The purpose of a tracer round is to enable a gun commander or gunner to watch the path of the fire and correct it onto the target. This is useful with distant land or sea targets and vital with airborne targets. Tracer can also be fired from a rifle and used to guide fire from other weapons onto an indistinct target. If the last round in your mag is a tracer you know when you're empty.

## Skill: How do you use the weapon to maximum efficiency?

Because an infantry machine gun may fire between 650 and 1,200 rounds per minute it is like a hosepipe for bullets. These bullets do not all go to the same place, as you might think,

### The ideal machine gun for infantry

Guns of 7.62mm calibre are ideal for use by and against infantry. The round is powerful enough and heavy enough to have a useful range of over 1,000m as an area weapon while it is still light enough to be transportable by men on foot.

As a general rule, the lighter the calibre the faster the gun can be made to fire – the old .50in Browning would be looking at 500 rounds per minute and heavier guns slower still. The type of machine guns used by infantry today fire from 650 to 1,200 rounds per minute depending on their gas setting and temperature. If you fire more rounds the gun gets warmer and fires a little faster. Of course, electrically operated heavy machine guns fire far faster – up to 16,000 rounds a minute in some cases.

A DShK 12.7mm heavy machine gun in a bunker at a FOB in the upper Kunar Valley of Afghanistan's Kunar Province, near the border with Pakistan. Although clearly in ISAF hands in this image, it is a machine gun that frequently features in Taliban arsenals throughout the region. (Corbis)

rather they are caused by the design of the barrel to spread into what is called a 'beaten zone'. This is an area of ground which receives the fall of shot from a machine gun firing on one bearing. At a range of some hundreds of yards the beaten zone is many yards long and wide but varies, of course, depending on the weapon and the age of the barrel. Machine gun barrels do wear out quickly.

## Types of machine guns

The British are fond of the 7.62mm x 51mm GPMG as it will fire up to 1,000 rounds a minute, hits hard and is very reliable. The US used the M60, which is quite similar in specification but slightly less reliable, for many years but have now replaced it with the M240 which is lighter, much more reliable and also fires the 7.62mm x 51mm round. Since the end of World War II the Russians have issued their infantry with the PKM machine gun which fires the 7.62mm x 54mm round – the 7.62 x 39mm AK47 round. Like all Russian weapons it is rugged and utterly reliable even when badly maintained.

Heavier machine guns you are likely to come across are mostly 12.7mm calibre which is the same as the .50in round. The West has used the M2 Browning .50 since World War II.

The reason being, I believe, there is little room for improvement in a weapon which rarely goes wrong and which knocks out 600 .50in bullets per minute each travelling at 887 metres per second. It is often referred to as the Fifty Calibre or 'Browning Five Oh', despite the cartridge being properly designated the 12.7mm x 99mm, as there is an identical but smaller version chambered for the .30in round. The Russians issued a 12.7mm machine gun called the DShK 1938 which comes with wheels and has an armoured shield option. It has been copied under licence by China and other countries and appears in insurgents' armouries all over the world. This old gun is slightly less powerful, firing a 12.7mm x 108mm cartridge, with a muzzle velocity of 850 metres per second, but should not be underestimated. Two of these guns successfully shot down a British Lynx helicopter in Northern Ireland in 1990. It has now been replaced by the NSV and Kord — both of which are much lighter and more accurate.

## Mini gun

This lovely weapon is properly called the High Speed Rotary Gun but often referred to as a mini gun, Gatling gun or roller cannon. In principle it is just a smaller calibre version of the roller cannons you find in many aircraft. It is electrically driven and usually set to fire 3,000 7.62mm x 51mm rounds a minute. So that is four GPMG or M240s! Obviously, due to the round it fires, the range and hitting power are very similar to the GPMG or M240. It is superb for area suppression and for returning a firm response when ambushed.

A US soldier holds a mini gun multi-barrel machine gun mounted on a Black Hawk helicopter in the Arghandab valley in Kandahar province, southern Afghanistan, 20 February 2010. (USMC)

Heckler & Koch 40mm Grenade Machine Gun mounted on a Jackal pictured in Afghanistan in 2010. (Photo courtesy Tom Blakey)

Some things never change – a piece of kit as familiar to an old-timer like me as it is to you today – a GPMG takes pride of place in this photograph. This particular GPMG was used in a sustained fire role in Radfan, Yemen, 1964. (Photo courtesy Colour Sergeant Trevor 'Sadie' Sadler, 1st Battalion 'The Vikings' Royal Anglian Regiment)

The problem is that this gun needs an electric power supply and weighs 66lb without its mount or ammunition. You don't want to carry one very far. And the recoil is 300lb of pressure against the direction of fire so don't believe anyone fires one of these from the hip either. For all the above reasons the mini gun has been mounted on helicopters, other aircraft and riverboats since Vietnam times.

More recently a firm called Dillon has begun manufacturing an improved version of the basic design, known as the M134D. This has many upgrades in detail, resulting in decreased weight of the system (especially when using a titanium gun body), improved reliability and better handling and maintenance.

I have mentioned it here because they are beginning to be used on light armoured vehicles as their high rate of fire makes them a comfort in an ambush situation. I think the enemy may find them a little off-putting as well.

### Heckler & Koch 40mm Grenade Machine Gun (GMG)

Effectively this is a machine gun which fires explosive shells. The shells will penetrate 2in of armour or any sort of wall and the detonation will kill anyone within about 5 yards and wound out to 15 yards.

The **GMG** or *Granatmaschinengewehr* as the Krauts call it, was developed by the fine German engineers at Heckler & Koch for the German Army in about 1992. It has been adopted by British and other European forces because it is truly excellent. There is a very similar weapon built in the US called the M19. The only significant difference is that the GMG can be set up to feed from either side and this would allow it to be mounted in a twin side-by-side situation.

Operation is blowback and rate of fire is 350 rounds per minute cyclic but this cannot be maintained in the field owing to the difficulty of feeding such a bulk of ammunition. A realistic rate of fire allowing for switching boxes and joining belts is around 40 to 60 rounds per minute.

The extreme range is about 2,200m but it is usefully accurate out to about 1,500. You can get all sorts of clever night sights to fit it too. In a word – excellent.

The weapon itself weighs about 64lb and the stand another 22lb. The ammo weighs a ton so obviously it is a vehicle mount by choice but you can dismount one to set it up on a perimeter. The recoil is not excessive, surprisingly, so the little tripod just digs into the ground nicely and gives a good firing platform.

## Summary: What are the important things to remember about machine guns?

Don't carry a machine gun on a foot patrol without considering the situation as the automatic rifle is often sufficient for firepower. Do try to obtain a machine gun when you are in a static position or have vehicles as then the advantage is free of the weight cost. A 7.62mm gun is plenty for use against infantry but a 12.7mm is very good for demolishing a building sheltering insurgents. Try not to get lumbered with carrying the gun. The GMG is also superb for taking out enemy sheltering in buildings.

# ROCKET LAUNCHERS

'Rocket Launcher' is the popular name for a range of hand-held anti-tank weapons. Technically, these are all 'Recoilless rifles' as the propellant charge is ignited and burns completely in a tube or barrel open at both ends rather than in a rocket motor carried by the projectile. Nevertheless, rocket launcher is what they are popularly called.

## Goal: When is the right time to employ this weapon? What can it do?

There are a few hand-held rocket launchers around but only one is used by insurgents and this is called the RPG – short for rocket propelled grenade. The RGP7 is common and the newer RPG29 models are around here and there. It will penetrate 10.6in of steel armour up to a range of 1,100m and then self-destruct by a time fuse.

Hand-held rocket launchers aren't used much against tanks. It is, of course, what they were originally designed for but insurgents don't have many tanks for you to shoot at and ours are pretty much resistant to hand-held rocket launchers because of clever armour.

Soft-skinned and lightly armoured vehicles have no defence against the RPG7 so they are often used to open up an ambush by stopping the lead vehicle – so to speak. On the good side, the blast may pass through the vehicle without harming many people as light armour is easily penetrated. However, when armour forms a box it can have the effect of compressing the shock wave inside the crew compartment with unfortunate effects on the crew members. The answer to this is a type of spaced armour grill mounted on the outside of the armour which causes the rocket to detonate too far away from the main armour skin to penetrate it.

Tom models the RPG7, Afghanistan 2010. (Photo courtesy Tom Blakey)

There was a certain American officer of my acquaintance whose idea of tactics was to drive his soft skinned command vehicle – a Mercedes Unimog truck – into an enemy base and everyone fired outwards... I'm not saying he didn't have lots of guts but perhaps that was more courage than sense. Anyway, one day he did this and the opposition fired an RPG7 at the truck. The projectile hit the low tin-plate side which set off the shaped charge. The blast came forward through the side and cut off the leg of the young man operating the radio. It took off the leg close to the hip without doing any other damage – funny how these things happen – and the flash of the blast cauterized and sealed the wound. The young soldier – 18 I think he was – became an instructor and a walking warning as to the danger of RPG7s. (Best regards if you read this, buddy.) Interestingly enough, he was assigned to teach armoured warfare to potential targets of the RPG7. Strange how the military mind works...

Helicopters are a favourite target for the RPG7 where heavy machine guns can't be moved to the attack site and shoulder-launched anti-aircraft missiles are in short supply. Choppers take a lot of knocking down so the RPG7 is ideal if the attacker can actually hit the aircraft. The tactic often employed is to have a whole group of people fire at the same chopper from ahead or astern and hope for a hit.

Because of their light weight and knockdown punch rocket launchers are used mostly where an explosive charge needs to be fired by an infantryman. For a Special Forces operator it could be used for destroying vehicles, a house or a bunker or it could be aircraft on the ground or helicopters at very short range.

## History: How did it develop to be the way it is?

During World War I the need arose for anti-tank weapons, but grenades and mines were the only types in common use, aside from field artillery fired over open sights of course.

Between the wars and into World War II several types of light anti-tank weapon appeared. The British began with a grenade launched from a rifle and this developed into a special weapon called a PIAT (Projector, Infantry, Anti Tank), which used a spring, amazingly, to launch a type of grenade at the target. This was the first use of what is called the HEAT (High Explosive Anti Tank) principle where a charge flattens against armour and on detonation either pushes through it or sends a shock wave through it to break off a 'scab' on the inside. The semi molten scab then flies around inside the tank with obvious results...

The Germans developed the first recoilless rifle with the Panzerfaust 150. This was a primitive RPG and the Russians copied it to produce the RPG2 which became the RPG7.

The recoilless rifle principle was also employed by the USA in the 'Bazooka' during World War II. This was a pipe of some 3.5in diameter, operated from the shoulder, which launched a projectile comprising an armour-piercing charge. When initiated by a trigger, the charge attached to the rear of the projectile pushed the projectile forwards but, because the gases were allowed to escape to the rear freely, there was no recoil, just a strong rearward blast. This enabled a single soldier to fire the relatively heavy projectile a long way horizontally without requiring a mounting to withstand the recoil.

The bazooka was refined over the years by Western forces, passing through several incarnations, which included the 84mm Carl Gustav, but retained the same operating principle. In recent years hand-held rocket launchers have become smaller and lighter while still packing a fair punch.

The USA has opted for very light disposable anti-tank rocket launchers while the Russians, and therefore most insurgents, have stuck with the RPG group. The latest model, the RPG29, has knocked out Merkava, Challenger and Abrams main battle tanks in desert warfare.

## Operation: How does it work?

There are two clever principles which combine to make a modern hand-held rocket launcher work. The first is that the heavy projectile is driven forward by the detonation of a propellant charge which is also allowed to escape out of the rear of the launching tube. Those with the physics will see that this stops any recoil. Those without physics will have to try firing one.

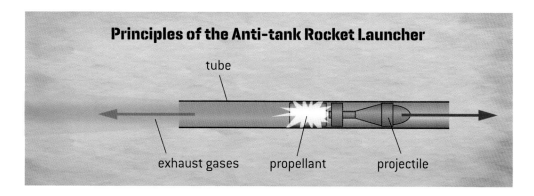

**Principles of the Anti-tank Rocket Launcher**

tube

exhaust gases        propellant        projectile

The other clever idea is the shaped charge. All rocket launchers employ the shaped charge principle in at least some of their projectiles. This enables a relatively small charge to generate an immense directional blast sufficient to cut through armoured steel of great thickness, often over 10in. I haven't supplied a diagram detailing the operation of a shaped charge on the grounds that its principles may be used by terrorists and insurgents. There are other types of charge which have been developed to defeat composite armour but we don't need to share that information here either.

## Skill: How do you use the weapon to maximum efficiency?

Using an RPG, or any other model rocket launcher, is not rocket science (sorry). Point and shoot but remember that you have a slow moving projectile compared with a bullet so when aiming at a moving target such as vehicles or aircraft try to get yourself in line with its direction of travel – either straight ahead or to the rear of the target so it is not moving across your field of view. Again, for obvious reasons I am not about to point out the weak points on our kit but be sure to familiarize yourself with these once on the ground.

## Types of grenade launchers

The Russians came up with what is probably the best light anti-tank weapon ever made. The famous RPG7 or Rocket Propelled Grenade Mk7. This employs the same principle as the Bazooka, Carl Gustav and LAW66 but the tube is narrow and the fatter projectile sticks out of the front of the tube while the candle-like propellant charge screws to the back of the projectile and then is slid inside the tube as the weapon is prepared for use.

Like all Russian weapons the RPG works every time even when dirty. It has a range of some 1,100m and then the charge self destructs – so if it lands on soft ground next to you

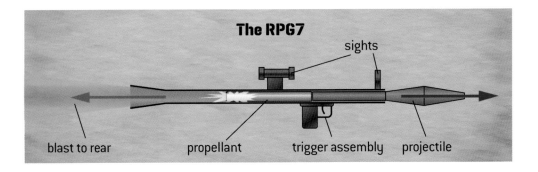

**The RPG7**

sights

blast to rear          propellant          trigger assembly          projectile

it will still go off. It is very accurate and has a similar penetrative ability to the Western weapons – some 10–12in of armour.

**The LAW66:** The USA has produced a miniature, lightweight disposable weapon called the LAW66 presumably from it having a 66mm calibre and being a Light Anti-Tank Weapon. This clever little toy uses the same principles as the RPG but is made in high-tech US factories so it weighs only a few pounds and telescopes up from around 18in to a working length of a couple of feet.

At close range it is effective against armour up to nearly a foot thick. I have myself tested them against sections of the HMS *Ark Royal* aircraft carrier armoured deck (after it had been scrapped you'll be pleased to know) up-ended on a firing range. The projectile cut a clean 66mm hole through 10in of armoured steel with only a few drips of molten steel to spoil a perfectly clean hole. I was quite amazed at the power of such a little thing. On the down side, the 66 can be inaccurate and the projectile has a tendency to bounce off armour if it hits at a shallow angle.

**The Wombat:** The Wombat was effectively a huge Recoilless Rifle with a 120mm calibre tube which was issued to the British Army some years ago. Several weapons of this type were in use around the world as medium-weight vehicle-portable anti-tank weapons. Very often you would see a Wombat-type weapon towed behind or mounted on the back of a Land Rover or other vehicle.

**FGM-148 Javelin Anti-Tank Guided Missile Launcher:** Rocket launchers have now been replaced by man portable missiles in NATO armies as these pack a bigger punch over a greater range and can often be guided onto the target by wire. This weapon is an anti-tank missile made by the US and is not to be confused with the Javelin surface-to-air missile built by the British. To maintain the confusion, the US and the British both use the FGM-148 Javelin as an anti-tank and anti-building weapon in Afghanistan and elsewhere. I shall refer to the FGM-148 as the 'Javelin' for convenience in this section.

Though designed to attack armour, the main use you are likely to make of this weapon is to destroy buildings or bunkers with enemy in them or knock down substantial walls. At around $32,000 (£20,000) a shot you don't want to waste them – unless, of course, you

FGM-148 Javelin Anti-Tank Guided Missile Launcher Afghanistan 2010. (Photo courtesy Tom Blakey)

are not paying. In any event the chances of you missing at anything up to 2,500 yards are minimal as you have excellent infra-red sights and are usually aiming at something the size of a house.

The Javelin has two modes of fire settings which cause the missile to travel either high or low to the target. These are Top Attack and Direct Attack. In Top Attack the missile gains height to around 500 feet and then drops onto the target from above on the premise that the armour on a tank is thinner on top. In Direct Attack mode the missile only climbs to around 200 feet and therefore travels on a flatter trajectory to the target. You would generally select Direct Attack mode to destroy buildings etc.

The Javelin is a 'fire and forget' guided weapon so the way it works is you line up on the target, select the mode of fire and let it go. The missile then guides itself onto the target and gives it a hefty thump with an 18.5lb explosive charge. When you press the trigger one propellant charge drives the missile out of the tube and there is some back blast. When the missile has travelled a little way the main rocket motor ignites and propels the missile the remainder of the way to the target.

The warhead is very clever and consists of two shaped charges which work together to defeat all the latest types of spaced, ceramic and explosive armour. During the invasion of Iraq these missiles proved devastating against the Russian T72s fielded by the opposition.

## Summary: What are the important things to remember about grenade launchers?

If you forget everything else I tell you, remember there is one hell of a back-blast from all types of rocket launcher. The choice is between the accuracy of the RPG7 and the light weight of the 66. On balance the RPG isn't so heavy and the tube – which I think is fibreglass or tin depending on the model – can be carried separate from the projectiles and propellant. I think the RPG just gets the vote for accuracy but the 66 is certainly better to carry. So it depends on the job and the circumstances – like so much in life. By the time the rocket is fired at you it is too late to do anything about it so make sure your vehicles are fitted with an external grid of some kind to detonate rockets and keep a screen of infantry between aircraft and potential RPG7 launch points. Maintain constant vigilance on convoy duty and use infantry screens as explained elsewhere.

# MORTARS

A mortar is essentially a light, simple artillery piece which is restricted to the indirect fire role. That means you lob a bomb up into the air to drop **onto** the target rather than being able to shoot flat over open sights **at** the target. Despite this limitation the mortar can be devastatingly effective when used properly at the right time against the right target. Or it can be a waste of time and effort when used incorrectly.

## Goal: When is the right time to employ this weapon? What can it do?

The best time to use mortars is against targets which are out in the open – perhaps troops on patrol or a vehicle convoy. In this case a concentrated barrage by a number of heavy mortars can do terrific execution. Where mortars come into their own is when a battery of three or more are carried on trucks or armoured personnel carriers (APCs). This means weight is less of a consideration so lots of ammunition can be carried. This kind of concentrated assault can have a very depressing effect on the enemy.

The standard procedure is as follows: mortar crews set up their emplacement – dug in safely – and use various techniques to aim at targets they usually cannot see. Ranging methods

A US Marine trains on an 81mm mortar. (USMC)

include surveying techniques and Global Positioning equipment. Suffice it to say that when they are given a target by grid reference they can drop a great many bombs onto it in just a few seconds.

Some miles from the mortar battery a forward observation officer spots a group of enemy. He works out the exact position of the enemy and tells the mortar battery where they are. (Just coming into service is some clever kit with a point and shoot laser that sends the information directly to the mortars so you may have the benefit of using this). A short while later a concentrated rain of bombs comes down on the enemy with extreme prejudice. In the ideal situation they have no chance to hide because the mortar bombs come from behind a hill or landscape feature that hid the sound of their firing and so they are all destroyed in the open.

## History: How did it develop to be the way it is?

Mortar tubes – the firing barrel – are pretty simple creatures and have been in use for as long as there has been artillery. What has changed over the years is simply that the tube has got lighter, the bombs more effective and the aiming gear more accurate. Mortars come in many sizes and are usually designated by their bore diameter or calibre. The smallest are the 2in/50mm and the 60mm with the next sizes up being the 81mm and the 82mm. The British and Canadians brought out the L16 81mm mortar and the US produced another 81mm called the M252 and both have been very successful. These strip down into three packs of 25lb/11kg each: base-plate, bipod and sites and tube.

The Chinese and Russians have both produced a 82mm mortar which had similar useful range and firepower as the first models used during World War II, used during the battles for Kursk, Stalingrad and Moscow. The 81mm cannot, of course, fire any captured 82mm ammunition but the Chinese 82mm variant can fire any captured 81mm ammunition. Damn clever those Chinese.

## Operation: How does it work?

A mortar is a tube which is fired at an angle steeper than around 45° from vertical. In use a finned 'bomb' is dropped into the muzzle. The bomb has propellant charges in varied amounts placed around the stem of the fins so as to achieve different ranges combined with the angle of fire. When the bomb either hits a firing pin at the bottom of the tube, or the pin is projected by means of a trigger, the propellant charges are ignited forcing the bomb out of the mouth of the tube in the same way as a shell comes out of a cannon but at a steeper angle and lower velocity.

The advantages of mortars include relatively low weight for transportation – certainly a tiny fraction of an equivalent artillery piece – and a high rate of fire, often as fast as bombs can be dropped down the tube, which is much faster than artillery of a similar calibre.

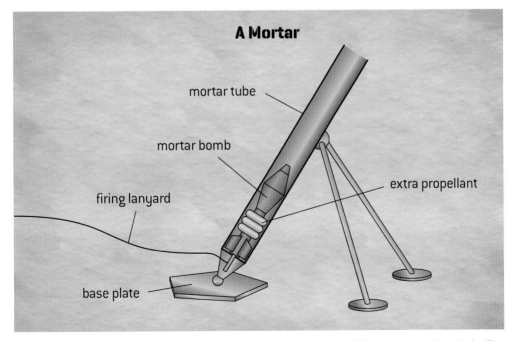

**A Mortar**

mortar tube

mortar bomb

extra propellant

firing lanyard

base plate

A diagram of a typical mortar as used by armed forces around the world. The extra propellant looks like a tea bag.

Disadvantages include a lesser range than artillery – though still some miles with larger mortars – and that they can only fire in what is called the 'indirect' fashion. This means that where a cannon – say a tank main gun – can either shoot straight at the target like a rifle or up in the air and down onto the target like a Howitzer cannon, a mortar must always shoot up into the air and drop the bomb onto the target. Indirect fire has a greater range for the same muzzle velocity and can lob a heavy shell or bomb further but the approach is slow and can often warn the enemy. Soldiers soon learn to recognize and discriminate between the sounds of different mortars being fired and also come to know how long they have to get under cover. Firing from a distance a mortar bomb can travel for many seconds. Plenty of time to find a trench or break-up a packed formation.

**Bomb damage:** When a mortar shell comes down it explodes as it hits the ground – ground burst – and the burst together with the shrapnel goes up and out. If you are laid flat only a few yards from a mortar you will survive as the blast will go over you. If you are standing you will die. To catch you in a trench the bomb will actually have to land in the trench with you. This is extremely unlikely as the diagram in the artillery section shows.

If mortars might be used against you make sure you don't dig trenches under, or spend any time under, trees. The reason being that the mortar bombs explode when they hit the branches of the trees above you and the blast comes down directly onto you so even a trench is not safe unless it has a solid roof. When terrorists use mortars they tend to use them in a primitive fashion without the accurate aiming or the massed fire of a battery. In my experience they will fire off a round or two at a building or patrol and then disappear. The effect is more a nuisance than anything else so don't lose your head when you experience this for the first time.

In one country I was deployed to in Africa we used to be attacked every night at exactly 18:30 when it became dark so quickly it was like someone had flicked the light off. The opposition used to regularly turn up just before dark and get just close enough to sight our camp then let off a few rounds at us and run off into the night knowing we couldn't follow them. We became very slack about this over time and everyone used to go to their slit trenches just five minutes early and have a smoke while the bombs came in. There were never any casualties despite scores of rounds – except once when I was cooking a fillet of gazelle over a fire suspended on a steel trip flare stake when I lost track of time. I got to my hole but the fillet was hit. There are no prizes for making life difficult in combat.

## Skill: How do you use the weapon to maximum efficiency?

The light mortars of 60mm sometimes issued to infantry are a waste of time as they are heavy to carry and have no useful effect owing to inaccuracy, lack of punch, lack of range and the inability to carry sufficient ammunition. The 81mm mortar used by Western forces and the 82mm used by Russian, Chinese and insurgents are much more useful with a range of 5.5km and a rate of fire of 12–20 rounds per minute per tube. The 120mm used by various nations are quite excellent weapons with a range over 7km and a killing radius of 70m from each bomb burst. As I said earlier, the odd shell is just a nuisance as the enemy can hear it coming but if you pick your time and catch a convoy or infantry formation out in the open you can destroy them. This is particularly useful if you are being attacked by large numbers of the enemy across open ground.

Dig in the mortars, as described later, and sight them in onto potential target areas. You don't get to play with heavy mortars without doing the course and there is quite a lot to it so that is probably enough for here.

## Types of mortars

By far the most common mortars are the 81mm used by NATO and the 82mm used by the Ex-Soviet Bloc and the Chinese. There are various 120mm mortars but in my opinion the 81/82 has the better balance of mobility and firepower. Just being issued for 120mm mortars in the US are bombs with a clever radar proximity fuse which, without being carefully set for timing, causes the bomb to explode at a set height above the ground. Very good kit if you have them as air burst is devastating against infantry in the open or in trenches. Rotten have used against you.

> ## Summary: What are the important things to remember about mortars?
>
> Mortars are lightweight artillery. Small calibre mortars of less than 80mm are a waste of time because they achieve little in return for the disadvantage of the infantry being forced to carry them. Heavy mortars carried by vehicles and used in batteries can do good work when used against enemy in the open or in a tree covered area for best effect. Mortar fire should begin as a heavy barrage to deny the enemy chance to react.

# LANDMINES

Because mines, and other explosive devices often referred to as IEDs (Improvised Exploding Devices), are a major problem to our forces in a counter-insurgency situation I have covered their use in more depth later. Here we will just look at how shop-bought mines work.

Landmines have had a very bad press in recent years. They are portrayed as awful things which maim and kill civilians and soldiers alike without discrimination several years after they have been placed and with no way of getting back at the people who planted them. Well, yes, in an insurgency situation they are all this – if you are the one treading on them. From the user's point of view they are clever little things. Scatter a few around and you deny the enemy the use of the land indefinitely. This is why they are so popular with insurgents.

Mines have an entirely different use in conventional warfare and I think it is important you understand this so I do explain it later. Having said that, in conventional warfare engineers lay mines so you may never touch one except with your foot. The mines you are sure to come across will be used against you by terrorists. So perhaps it is even more important you understand these.

## Goal: When is the right time to employ this weapon? What can it do?

A landmine is a defensive weapon in conventional warfare and should be laid in a clearly marked area around your position to funnel or slow the enemy as they remove the mines under your fire. As I said above, this type of minefield is generally laid by engineers but if you were an attacking force in a conventional situation, or perhaps on a Special Forces operation, you might come across this type of minefield. Of course, that is not how you will come across them when you are trying to put down an insurgency. In anti-insurgency warfare the mine is used

## REMEMBER:

Anti-personnel (AP) mines will only burst a tyre on a vehicle and there is little collateral damage especially if the tyre is full of water (the water is believed to help absorb some of the energy from the mine). This is another good reason to ride rather than walk. The greater danger will come from an ambush after the AP mine has exploded.

by the insurgent — scattered in ones and twos without marker or warning — to kill or maim both civilians and the forces of occupation.

There are two main types of mine. The anti-personnel (AP) mine which is used to cut off a person's foot, so they lower morale with their complaints and require casevac, and the anti-tank mine which is designed to take the wheel or track off a vehicle to disable it. The use of a mine against vehicles in conventional warfare may have several beneficial outcomes: the user can either hope to kill occupants of the vehicle, deny the use of the vehicle to the enemy by damage or disable it to make it an easy target for their guns. In a counter-insurgency situation a mine is often used to stop the lead vehicle in a convoy before the ambush party opens fire on a stationary target.

## History: How did it develop to be the way it is?

The mine started out as a bonfire placed at the end of a tunnel dug under the enemy castle wall. When the fire was lit the tunnel supports burned and the tunnel came down bringing the wall with it. One reason for moats was to stop this 'undermining' of fortifications. Later the fire was replaced by an explosive charge with more effective results. In the last century mines triggered by foot or vehicle pressure were developed by all the major players and became more effective as technology improved. They were considered an excellent deterrent to a mass infantry or tank assault on a position. They still are.

## Operation: How does it work?

Pretty much all the landmines you will come across are triggered by the pressure of a foot or a wheel or a track. There are some less common mines which are triggered by a trip wire and set off their charge above ground for a better area of effect. Some anti-tank mines are made to wait for the second pressure to catch the second road wheel of a tank and achieve more damage. There are other explosive devices also referred to as mines such as the Claymore which we will look at elsewhere as the principle upon which they operate is entirely different.

**Blast Mines:** AP mines come in a thousand different patterns but the vast majority are 'Blast Mines' designed to take the foot off anyone who treads on them. They achieve this by sitting in the ground waiting for someone to either tread on a pressure plate in their lid or knock a sensitive spike coming out of the top. When the mine is activated the charge explodes and the resultant shock wave takes off the pedestrian's toes, foot or leg up to the knee depending on how it was trodden on, the type of mine, how deep it was laid and the type of soil it was laid in. Gravel or stones are the worst as they provide natural shrapnel.

**Jumping Mines:** You will understand that the Blast Mine disables one man. A Jumping or Shrapnel Mine is intended to get some of his mates as well. Again there are a number of types but the principle is that when one man treads on the mine it waits a couple of seconds until he has moved on and then jumps up into the air where it is caused to explode either by reaching the end of a tether wire or a time fuse. The jumping part of the mine tends to have a larger charge and its own shrapnel casing. This can take out a number of men in one shot. Though they are not in common use they are quite unpleasant things to come across.

**Mine detection:** We look at this in more detail later, but in principle you find a mine by touch with a foot or, preferably, bayonet or by using an electronic mine detector of some kind. To counter this some mines have anti-lift devices set to explode on movement and many are made with as little metal in them as possible to make them hard to find.

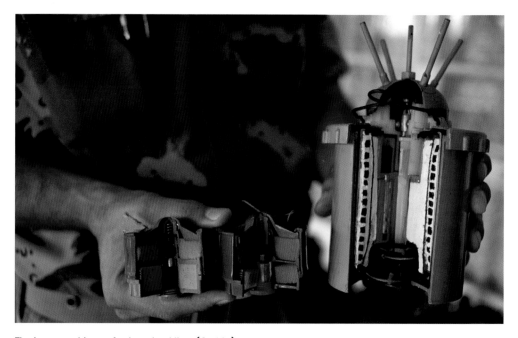

The inner workings of a Jumping Mine. (Corbis)

**Anti-tank mines:** In principle, anti-tank mines are similar to AP mines but have a bigger charge to defeat armour. Shop bought anti-tank mines take a lot of pressure to set them off so you could, theoretically, walk over them safely while they wait for a vehicle. Anti-tank mines will often roll a light armoured vehicle over but they are survivable. We look at their use later but for now be aware that insurgents often use anti-tank mines with AP mines laid on top so they are set off by a man and take out anyone standing nearby. Watch your spacing.

## Skill: How do you use the weapon to maximum efficiency?

In conventional warfare mines are laid out by engineers in certain patterns and areas to slow, funnel or deny access to attacking forces. I have covered this in a later chapter. If you are going to use mines in a counter-insurgency or false flag role when working as Special Forces then see the section later where I show how they are used.

Moldavia National Guards surface-lay Russian-type anti-tank mines on a bridge to prevent tanks crossing the Dniester River – a contested piece of land following the collapse of the Soviet Union. Russian anti-tank mines have been used against ISAF forces in Afghanistan but increasingly IEDs are the primary threat. (Topfoto)

## Types of landmines

There are hundreds of types of both anti-personnel and anti-tank mine available all over the world. These come in all shapes, sizes and with varying degrees of effort built in to make them difficult to find and/or move. They are all pretty much the same from the principle of use and avoidance as we will see shortly.

Various types of AP mine. The one on the far right sits above ground with a trip wire and gives everyone a piece. (Corbis)

## Summary: What are the important things to remember about mines?

In an effort to make sense in this book I have shown you how shop-bought mines work here. Later I will explain the tactics used with and against both mines and IEDs together. This is because in counter-insurgency warfare the use of mines overlaps with IEDs/home-made bombs.

# PERSONAL EXPLOSIVE DEVICES

There is a whole range of explosive toys which you will come across as a soldier and, you will be pleased to hear, there are even more available to the Special Forces operator. In this chapter I will give you an overview of the most common so you know what they are and how they can be used for your benefit. The devices which I am discussing here are the shop-bought items with which you may well be issued. They fall into two main categories: lighting up the enemy and killing him.

## Lighting up the enemy

With the development of night-vision technology in the last couple of decades the use of flares of various kinds might seem to be on the wane. To a great extent this is probably true but for the sake of a little ink you might as well know what can be achieved by you or the opposition in this area. Remember the opposition might not be so well equipped with night-vision equipment as you are and might decide to light up the night so they can see to shoot you. If you are lit up by some incendiary device at night it is often better to freeze than move because the light they give off is of a type which makes colour vision, and therefore identification of a stationary object, difficult but allows movement to be spotted easily. If you find yourself in a situation where the enemy have night-vision equipment and you do not then frankly it is time to take your ball and go home.

SS 'Totenkopf' troops go to ground and freeze as a Russian flare bursts over a patrol in Demyansk, 1942. The principles of how to use flares on the battlefield haven't fundamentally changed in 70 years. (Cody)

**Rockets:** When you think the enemy are approaching you can send up a type of rocket, which will burst in the air like a firework and light up the whole area for a matter of some minutes as it descends by parachute. Certainly for long enough to make anyone approaching wish they were elsewhere. The benefit of rockets is that they do not have to be set up in advance at the site of the required illumination but may be deployed over an area which you have neither visited nor control. A disadvantage, compared to night sights, is that the enemy have a good idea that you have seen them coming.

**Ground flares:** As an alternative to rockets, when you have the opportunity to set them up along those lines of advance which you expect the enemy may use, are ground flares. These are stubby cylinders the size of baked bean cans which may be attached to a metal stake and set in the ground. They are set off by either a pull on a cord or battery and wire.

By far the most useful feature of a ground flare is not really the illumination. It is that it can be set up, say, outside your base perimeter, with a gun covering it and the pull cord to set it off can be used as a trip wire. This way when the enemy approach they warn you by setting off the flare and then you have light to shoot them by. A tireless sentry which is economical and convenient as they weigh only a few ounces. Not as good as a Claymore though.

## Neutralizing the enemy

There is a very popular explosive device which is much spoken of by those who have never used one. This is the Claymore Mine. It is not a 'mine' in the popular meaning of pressure-detonated landmine. Rather it is a type of shaped charge which is mounted above the ground on small, wire legs and detonated by wire and battery. I will speak of it at length here it is very useful and there is a whole range of similar devices which you will then understand.

The Claymore is an explosive charge shaped in such a way that the greater part of the blast is directed to propel shrapnel horizontally in a chosen direction. Claymore – as in the Scottish heavy cutting sword – is an appropriate name for this weapon as it cuts straight through any bodies within its range and the original was invented, appropriately, by a Scotsman named McLeod.

Picture a segment of an orange made of explosive only less curved. Imagine that the part which would have been next to the peel has a layer of shrapnel embedded in it. Then imagine the whole is wrapped in a waterproof casing with a hole in the top for a detonator and holes at either end, underneath, for the insertion of wire legs. A Claymore looks nothing like this but the idea is sound.

## TOP TIP!

### How to use Claymores effectively

To set up a Claymore pick a site where the mine can stand and point along a track towards oncoming enemy pedestrians. Remember there is a severe back-blast from a Claymore.

Instructor arms a Claymore mine. If this were not a demo with a plastic mine there would be no one else close by. Trust me on that. (USMC)

The effect of a Claymore is somewhat like a hundred shotguns all firing together in the same direction. The blast and shrapnel will cut a wedge something like 30° wide from floor level rising at something like 20° and, depending on the model, kill anything out to over 100m. These figures are not exact but simply based on the countless times I have used them. If all else fails read the handbook.

Open up the two pairs of wire legs and fix them into the base. Stand the mine where you want it to go remembering to point it the right way. It does say on the cover which side you point at the target area. Indeed, on the original mine, it said, 'Front towards enemy' for the hard of thinking. Try to get it right every time.

You can either set up the mine to be triggered by a trip wire or by a command detonation where you make the battery contact yourself.

**Whatever you do remember this next bit:** Lay out the wire from your firing position to the mine. At each end of the double wire, twist the two strands together to earth any residual static charge as this may be enough to set off a detonator. I once saw a detonator set off because this was not done. Fortunately it was not in a charge at the time but it frightened me half to death and earned the negligent person a sound beating.

Next attach the detonator by its two wires to the two (untwisted) ends of one end of the double wire. Always pick up a detonator by its wire. Do not ever touch a detonator body with your fingers. Body heat WILL set them off. Detonators are often referred to as 'Two finger', 'Three finger' and 'Four Finger' detonators for morbid reasons relating to how many fingers a particular size of detonator will remove cleanly.

When the detonator is attached to the wire, and you are ready to arm the Claymore, insert the detonator into the hole in the top of the casing. Retire briskly. The Claymore is fired by simply untwisting the wires at the command end and touching them onto a battery. If you do not use the Claymore and the urgency passes, you may want to keep it in position ready for the next night. Remember to take out the detonator at the first opportunity. Generally this will be done by the clearing patrols detailed later.

The safety tips for arming the Claymore pretty much apply to the use of all command detonated explosive devices with which you will come into contact. The thing about explosives is that your first mistake does not generally have the chance to make you careful as it is usually your last.

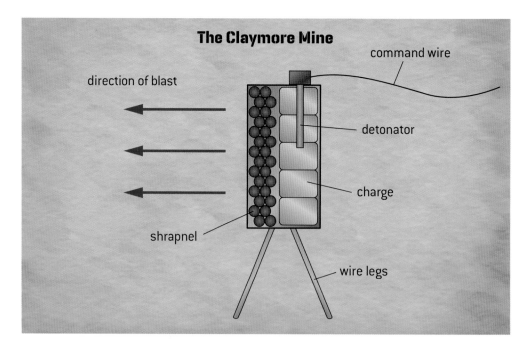

**The Claymore Mine**

command wire

direction of blast

detonator

charge

shrapnel

wire legs

Corporal Ndlovu prepares the gazelle skin after the Claymore had done its damage – not sure what for as it was full of holes. (Author's Collection)

I have often used the Claymore when defending small positions out in the 'sticks'. Claymores are very light to carry, pack an immense punch and can act as a tireless sentry meaning extra sleep is an option. You just have to be very, very careful with them. It is all too easy to get careless when you are doing this procedure every day and are bone tired.

The procedure I adopt is to set up a Claymore pointing down each of the main approach tracks the enemy might use when creeping up on our position. And then attach them to a trip wire trigger. The mine would then be 'covered' by a machine gun as a gun should be placed in the position of expected approach in any event.

One night I was dosing behind the gun in my shell scrape when the Claymore to my front went off. It was pitch black, and the sentries had lost their night vision from the flash, but I loosed off 500 rounds into the area. There was no reply so I stopped firing. The next morning there was a small gazelle which had been well tenderized by both mine and gun. At least the poor little thing would have felt nothing. And it was not wasted. Actually, it was this same little creature I was cooking on a trip flare stake when mortared as mentioned previously.

## Use of explosives in demolition

All armies have a demolitions course for soldiers who may need it. It is usually run by engineers who do this sort of thing all the time. The SAS and some other Special Forces units send all their men on a demolitions course as it is a useful skill if you are intent on disrupting an enemy's lifestyle with very few men. You will doubtless understand why I am not about to teach you here how to demolish an office block with a can of oil or produce explosives in bulk from household materials or design cutting charges or place charges for maximum disruptive effect. Pass selection for Special Forces and you can do a proper demolition course.

> ## Summary: What are the important things to remember about personal explosives?
> Never touch a detonator with your fingers — hold it by the wire. Flares and Claymores make really good sentries but watch out for the back-blast from Claymores. Don't get careless.

# SUB-MACHINE GUNS

Sub-machine guns (SMGs) fire 'automatic' pistol ammunition with a true automatic fire option and are fed from a magazine. They have a longer range and higher rate of fire than a pistol but cannot easily be hidden. They do not have the range, rate of fire or hitting power of an automatic rifle so their use is extremely limited. The only time I can see a use for a sub-machine gun is in a role where it needs to be silenced. You will be aware that the subsonic pistol can be silenced where the supersonic rifle cannot because of the sonic crack made by the rifle bullet.

## Goal: When is the right time to employ this weapon? What is it good for?

When would you want to fire a burst of pistol ammunition rather than a burst from a rifle? Are you a mafia enforcer wanting to shoot up another family's bar from your Sedan? I find it difficult to imagine where you would use a sub-machine gun in a military situation. I know

**ABOVE**
A silenced Heckler and Koch MP5 SD TRH215526 sub-machine gun. The 'MP' designation is used because the Krauts, quite rightly, consider them machine pistols. (Cody)

**LEFT**
Estonia: the SMG in this armoured case can be fired by a trigger in the handle. (Photo courtesy Sergeant Roy Mobsby)

many SF teams are issued with them for use in house clearing but I remain to be convinced they have any advantage over a shortened assault rifle. Perhaps the lack of penetration, compared to rifles, is why so many police carry them.

Probably the only reason for ordinary soldiers to be carrying a sub-machine gun is if they are in a vehicle and don't expect to use it. You might issue one to a clerk who was not practiced on a pistol so they might hit something. Hopefully the opposition. Tank crews used to carry sub-machine guns as they were short of room – there is surprisingly little room inside a tank. Feasibly air crew might do the same. Truck drivers do and shouldn't. They are giving up the advantage of a rifle without sufficient cause. Imagine your soft-skinned convoy is ambushed and you dismount and take cover then return fire. The enemy are 300m away blatting at you with AK47's which means they are out of the range of your sub-machine gun. Damn.

The US military has produced a version of the M16 rifle, which is 40in long, with a carbine version of 30in for when much of the firing is expected to be done from vehicles or inside buildings. This is just the ticket. Someone is thinking straight.

If you were on an operation where you needed to kill sentries quietly from a distance then a silenced sub-machine gun would give you better firepower and range than a pistol. Just to make sure everyone understands here, all weapons can be silenced for muzzle flash and 'thump' but a rifle round makes a loud supersonic crack travelling through the air and renders silencing futile. All pistol and SMG rounds can be loaded to be subsonic and therefore not make a supersonic crack.

## History: How did it develop to be the way it is?

The first sub-machine gun made its appearance at the end of World War I when rifles were bolt action and the rate of fire available was a distinct advantage. Between the wars various models were developed with the Thompson .45 being the most famous. This was fed from either a straight magazine or a drum capable of holding 50 rounds. By World War II with rifles still being mostly single shot, all the major armies developed sub-machine guns with the German Schmeisser, the British Sten and the US Grease gun being well known examples.

In Israel, Major Uziel Gal developed the Uzi sub-machine gun in the late 1940s and this heavy but small and efficient weapon became the standard by which all others were judged. At the present time there are a number of sub-machine guns in production with the emphasis being on small, light and high cyclic rate of fire. Some over 1,000 per minute. Glock actually produce an accessory for their pistols to convert into a sub-machine gun. I think this is more to cater for gung ho-amateurs than serious operators although the increased accurate range over a pistol, given by the stock, may give some advantage to bodyguards.

## Operation: How does it work?

All sub-machine guns are magazine fed and driven by the blowback principle where the pressure on the cartridge case to move backwards after firing pushes the breech back to extract the last case and feed the next. Many have folding stocks to make them even smaller. Almost all Western sub-machine guns fire either 9mm or .45in rimless pistol ammunition. They typically have a slightly longer barrel than a pistol which develops a little more power from the charge and a bigger magazine to carry more rounds. You already know what effect to expect from these rounds. Because sub-machine guns work like automatic pistols but fire many more rounds they are vulnerable to jamming through a build up of carbon on the working parts. So be sure to keep your sub-machine gun clean. Most sub-machine guns are fitted with a shoulder stock to steady the weapon and give better accuracy. But don't expect too much as you will be doing well to hit anything at much more than 100 yards.

## Skill: How do you use the weapon to maximum efficiency?

I was taught to double tap a couple of rounds at a time with an SMG and using this technique you can hit out to 100m without any trouble. But then you do have to wonder what the main advantage is in using an SMG as you are not using the firepower. Why are you not then using a rifle? Because of the stock and longer barrel you will achieve more accuracy than a pistol but so far as I can see the only time a sub-machine gun has anything over any other available weapon would be a rapid burst of fire from a team of bodyguards at close range to suppress an attack on a VIP. Or law enforcement using SMGs to overawe criminals carrying pistols.

As I said above, the time to use a sub-machine gun is when you have to sneak up on guards or similar and kill them quietly. Double tap into the centre of the body at distance or head if you are closer. (A double tap is where you squeeze and release the trigger so as to fire off two or three rounds in each burst. A little practice and you quickly get the feel of it.)

## Types of sub-machine guns

There are dozens of sub-machine guns available from the old Thompson .45 'Tommy Gun' through the Sterling 9mm and the Uzi to the modern Glock. I don't think any of them are worth carrying unless you need a silenced weapon. And then a pistol is often better.

## Summary: What are the important things to remember about sub-machine guns?

Because they make lots of noise and throw lots of lead around people tend to overestimate the effectiveness of sub-machine guns. It is better to think of them by the German name of 'Machine Pistol' because, effectively, that is what they are. They are longer range pistols with a large magazine and an automatic fire selector.

The sub-machine gun is neither fish nor fowl. You can't hide it like a pistol and it doesn't have any of the good points of a rifle – so I would always choose a rifle instead. Ideally only use it when it can be silenced, for shock value or when size/space is an issue. If you do use a sub-machine gun watch your ammunition use. Keep it very clean because sub-machine guns jam easily. If you are in a hurry, tuck the butt into the middle of your stomach and sight with both eyes open along the top of the gun.

# PISTOLS – REVOLVERS AND AUTOMATICS

I am going to shatter some dreams and fondly held beliefs here but I am trying to make you a better soldier and keep you alive rather than make a friend of you so think about what I have to say before you start pouting. I believe soldiers should not carry a pistol as a back-up weapon when in combat. The **only** time for you to carry a pistol is when you are working undercover in plain clothes.

## Why do so many soldiers carry a pistol in reserve?

The problems with pistols...

- Even the most powerful pistol is far less powerful than your rifle. It has to be or the recoil would break your wrist. The pistol bullet travels at a relatively slow speed and therefore does not cut up the target's innards like a rifle bullet.
- A pistol is very inaccurate compared to a rifle. If you doubt me look at pistol competitions and you will see they are all shot over very short ranges whereas rifles are long range.
- The pistol has a low rate of fire because it only has room for a small magazine.

The reason many soldiers give for carrying a pistol as a reserve weapon is in case their rifle jams. Actually it is because they have seen too much TV. If your rifle is clean and it jams you don't need a pistol you need a new rifle. Time after time I see gung-ho Private Military Contractors carrying pistols as a back-up weapon in case their rifle jams when actually carrying that pistol is a liability for two reasons. If you carry a pistol as a second weapon, as a soldier in an unfriendly crowd you are likely to get it snatched and used against you. In the USA about five police officers are shot and killed with their own weapons every year and the principle is quite similar.

If you carry a pistol and your rifle jams you are going to go for the pistol. Fine, I hear you say. Tell me this: if you are not Wyatt Earp, how long does it take you to get out your pistol, cock it or take off the safety and then begin shooting? Is it quicker than you can cock your rifle to clear the stoppage? I don't think so – and if it is you need to practice your stoppage drills. You need to practice them anyway until you do them without thinking.

Pistol training. British Army, 2010. (Mike Fletcher © UK Crown Copyright, 2010, MOD)

## So what are pistols good for?

Pistols are really useful for police and other people whose main job is to talk to people and tell them what to do rather than kill them. Police officers in many parts of the world carry a pistol all their lives and maybe never need to use it other than to intimidate people. In this situation, the weapon is there as an implied threat. 'Don't mess with me and do as I say because I have a gun.' In some parts of the world law-abiding civilians can carry pistols for personal protection. In this case the pistol is either a threat if visible or to kill an aggressor if not. Criminals, of course, in most parts of the world carry pistols for almost the same reasons — to make people do as they are told or to kill their enemies.

But what we are interested in here are people like undercover operators who need to carry a weapon which other people don't know about. A hidden weapon which allows them to mingle with civilians in a bar unnoticed yet kill efficiently at short range when they need to. The rest of this chapter is based on the idea you will use a pistol when you are working in an undercover role or on a covert mission as an SF operator to eliminate a bad guy. I say 'bad guy' because we only kill bad guys, of course. In this role, the firepower of a pistol is enough for what you need.

# Revolver or automatic?

There are two types of pistol, the revolver and the automatic* but as they fire roughly the same ammunition, the hitting power – or lack thereof – is about the same for each type given a similar round. The main trade-off between the revolver and the automatic is that the revolver never jams while the automatic rarely jams and has a higher rate of fire.

The key difference between automatic and revolver ammunition is that the automatic round is rimless – having a groove around the base to allow the extractor claw to get a grip on it and to stack better in a magazine while the revolver round is rimmed i.e. it has a projecting rim at the base to stop the round sliding through the cylinder.

So, there is nothing to choose in stopping power, accuracy or weight of weapon but the automatic has a greater rate of fire owing to the larger magazine and quicker magazine change while the revolver is not capable of jamming. While the higher rate of fire may sound attractive, the time you really need your pistol you will only need to fire a couple of rounds and you will certainly die if your weapon jams. I know several professionals who still carry revolvers when covert so the choice is not so clear as it may seem. On balance, and because of the work I do, I have come to prefer the automatic owing to the reliability of modern automatics and the higher rate of fire. But then I use a pistol regularly and I don't get very excited.

# The revolver

The revolver carries its ammunition in a cylinder magazine holding only five or six rounds and the cylinder rotates before each shot to bring the next round into line with the barrel. To reload the cylinder swings out of line with the barrel allowing access to the rear where the rounds are inserted. There are little gadgets available which allow all the rounds to be slipped in at once for speed.

Revolvers come in two kinds: single action and double action. Single action is the old-fashioned type where the hammer above the trigger has to be pulled back by hand to cock the weapon and bring the next round and chamber into line for firing. The cowboy-era pistols were of this kind hence gunfighters employing the technique of 'fanning' to achieve a higher rate of fire. Fanning is where the trigger is held depressed, or mechanically fixed so,

---

* The so called 'automatic' pistol is actually semi-automatic as the term 'automatic' means a weapon which fires cyclically and repeatedly for as long as the trigger is depressed – such as a machine gun. A semi-automatic fires once each time the trigger is pressed and then reloads – usually by recoil action pushing back the top slide in a pistol and gas action in a rifle – to be ready for the next shot.

and the hammer is repeatedly and quickly pulled back and released with the heal of the palm on the free hand.

Double-action revolvers use the energy from squeezing the trigger to bring the hammer back and turn the chamber to bring the next round into line. Effectively, therefore, the first part of pulling the trigger back cocks the weapon and the next bit of travel fires it. The advantage of double action is that the weapon can be fired quickly with one hand. The disadvantage is that the pressure required to cock the weapon with the trigger finger tends to pull the aim off to the side and leads to inaccuracy. All double action revolvers can be used like a single action – cock and fire. The best way to take advantage of the double action revolver is to cock with the thumb of the shooting hand when you have time before the first shot. There is a knack to this but it takes practice.

## The automatic pistol

The so-called automatic carries up to c.13 rimless rounds in a spring-loaded vertical magazine, usually in the hand grip, and the whole top slide of the weapon is driven back on firing to extract the empty shell case and compress a spring. Then the slide comes forward under spring pressure to pick up a round from the magazine and feed it into the chamber ready to be fired on the next squeeze of the trigger. When the magazine is empty the top slide stays back and reminds you to change magazines – which can be a help to a busy mind.

The automatic pistol has to be cocked by, usually, pulling back the top slide and releasing it to feed the first round into the chamber. Thereafter each time the trigger is pressed the pistol fires and the recoil pushes back the top slide to pick up another round from the magazine in the handle and feeds it into the chamber ready for the next shot. So in reality it is a semi-automatic weapon. I will continue to call them automatics because everyone else does.

The main advantage of the automatic is that it has a higher rate of fire than a revolver and can hold more rounds in the magazine. Also the barrel is not pulled off aim before each shot by the trigger pressure required to turn the cylinder. The magazine capacity is limited by the size of the hand-grip and the size of the rounds. There are 9mm automatics such as the Berretta and the Heckler & Koch which will hold 13 rounds while the Colt 45 will only hold seven rounds because they are so fat.

The original .45 automatic was designed for the US Army around 1911. The specification they asked for was a reliable pistol which would stop a charging horse at close range. Later variants are still in use today with the US Army because of their reliability. The only downside to the .45, within the limitations of it being a pistol, is that it is bulky to hide and very heavy.

## Common pistol calibres

Though all handguns have names and numbers, much like cars, they are often referred to by the calibre of the round they fire as this is what really counts. As with all firearms, you don't kill someone with a pistol. You kill them with a bullet. The calibre and velocity of the bullet determines its hitting power and also the cartridge which propels it. These two together form the round which determines the whole weapon's size, weight and utility. Just like with a rifle as we saw earlier. So, if we look at the rounds, the part doing the job, it will tell you most of what you need to know when making a choice amongst the thousands of types of handgun available.

I think the best way to get the range of calibres across to you is for me to go through the most common calibres of pistol round available in order of stopping power with an outline of the characteristics of the weapons which fire them. Then you can weigh up the pros and cons and make your own choice.

The .22in round for a pistol is very much lacking in power but it can be fired from a small light discrete weapon such as the Walther PPK automatic and it is quiet for a pistol too. There are many revolvers and pistols which fire .22 rounds so it is not difficult to obtain one and the ammo is common and cheap in the civilian market. To be sure of a kill with a .22 you need to go for a head shot so aim for the centre of the face. The main advantage of the .22 round is that the weapon which fires it is tiny and so easy to hide. Such a weapon is often carried by well-trained assassins and these may favour a hollow-point poisoned bullet.

The .38in round is normally fired from a revolver and is rather underpowered. Much better than the old .38, which is now no longer used very often, is the .38 Special which is the same calibre with a larger propellant charge in a longer cartridge case. It is actually half way between a .38 and .357 Magnum in terms of power. This is probably the ideal law enforcement revolver being light and convenient with adequate stopping power at close range.

The 9mm round is the favourite of the European armies for pistols and sub-machine guns and the Americans are beginning to use it too now that more powerful and effective ammunition has been developed. The 9mm needs a much bigger weapon than the .22 but is far more powerful. Like most things, it is a case of swings and roundabouts. In the past the

9mm did get a reputation for being a little underpowered but what I suggest is that you practice with the standard issue rounds and when you come to action buy yourself the Belgian Specials which have an armour piercing capability. 9mm pistols such as the Browning and the Star are cheap and common while the ammunition is easily available in most theatres. While you won't use much pistol ammunition in combat you will use lots in practice if you have any sense at all.

The .45in rimless is best known as the ammunition for the .45 Colt automatic mentioned above. You only get seven shots but the re-load is quick and if you can hit them they will go down. The only downside is that if you are on foot it is bulky and a lot of weight to carry. There are lots of old revolvers around which fire .45 rimmed ammunition but they too are too big and heavy to interest us here.

The .357in is well known as one of the 'Magnum' rounds. It is usually fired from a revolver although there are automatics available. Yes, it is quite powerful, being a sort of supercharged .38 Special, and will stop anything that walks but the weapon that fires this round is ridiculously heavy as it has to be to handle the mechanics of firing the round and steadying the recoil. I used to carry a long barrelled 'Blackhawk' .357in Magnum as my reserve weapon in a shoulder holster until I had to use it. My excuse is that I was young and wanted to 'look the part'. There is actually a .44in Magnum round but the cannon which fires it is too big to hide or carry very far.

Some military-philosopher guy once said that war is a series of mistakes and the side that makes the least mistakes wins. Veterans reading this will laugh because they will already have seen that most wars are a series of cock-ups. Just so you don't get the idea that I think I know everything, let me tell you a story where I was carrying a pistol as a reserve weapon and it did, sort of, come in useful. But when you have read the story ask yourself how often this is going to happen.

In Mozambique we had just cleared a base camp of some hundreds of enemy soldiers. We had helicopters with us to take out the wounded though we came in on foot. (I still have a soft spot for the old US Hueys born of gratitude.) The choppers were taking fire from the hills around the camp where there were a number of heavy machine guns sited – 12.7 to 22mm Russian weapons which would knock down an old chopper or at least make the pilot bad tempered. A certain US major sent me with nine guys to clear the top of one of the hills of anti-aircraft guns.

On hands and knees we climbed up this rock covered 'Gomol' as they call the hills there. When we got 30ft from the summit the guys dug in there started dropping hand grenades on us. An American lieutenant went quiet and we thought he was dead. A grenade landed about 3ft from my head and I just had time to think, 'Oh Shit', or similar, before it went off. I took 12 pieces of shrapnel down my right side from one which hit my right temple and slid along the bone above my eye to some in my thigh. One other piece somehow went through the sole of my left boot and into my foot. I must have had that leg crossed behind me – so the urban myth is untrue that lying with your feet towards a grenade will protect you.

The experience of being hit by shrapnel must vary according to where you are hit and your mental state at the time. Adrenaline numbs most things. For me it felt like what we used to call 'dead leg' at school – where someone knees you on the thigh and it makes the muscle go numb.

There turned out to be 600 of the opposition on the top of this hill. They were dug in to rock in tight trenches and it took a battalion of Special Forces later to get them out.

At this point I was quite tetchy because I thought I was going to be left there as it was getting dark and the team were being forced back down the hill. Fortunately it was just my right eye which was closed by the blast so I was able to sight my rifle with my left and knock off the heads which appeared over the crest. Trouble was, when I tried to move from the isolated rock I was hiding behind a hail of bullets drove me back to cover. A young Welshman named Taffy Davies was quite close and he provided covering fire to keep their heads down and suppress their fire as I moved from my rock. Taffy opened fire and I hobbled out for the next rock down the hill holding my pack in one hand and my rifle in the other. Out in the open my pack was shot out of my hand and then a round hit my rifle knocking that out of my hand as well and sending it spinning down the mountain. I kid you not. It didn't take me long to reach cover but I was breathing heavily.

Drawing my Blackhawk reserve revolver like John Wayne I lead the team back down the Gomol to our base camp in the dark, limping. Happily, I didn't have to test my aim with this heavy sidearm as I was getting woozy from loss of blood. I was casevaced out later and stitched up. The American lieutenant had taken a piece of shrapnel in his backside and we were pleased to see him limp in later. Not being a Hollywood star, I had chosen to leave the lieutenant and save the rest of the team rather than search under heavy fire in the dark. This was the last time I carried such a heavy reserve weapon, changing afterwards to the excellent Berretta 9mm.

## Pistol tactics

Civilians mostly get their knowledge of pistols from Hollywood. Police officers and cowboys dropping the bad guys with one shot from 100 yards. The reality just isn't like that and you will never kill anyone with a pistol at more than about 10 yards. More likely 5. Remember I said 'will not' rather than 'can not'. For that matter, TV and the movies often show lots of shooting with rifles, sub-machine guns and pistols at point blank range and no-one getting hit because of the laws relating to violence on TV in some places – reality ain't like that either. You will pretty much always use your pistol at close range in a building, car or possibly a back alley. If that were not the situation you would be carrying a rifle. Spook work is a lot seedier than it looks on TV.

Pistols are actually quite difficult to aim well because they are so short barrelled and most people need a lot of practice to get to be any good at hitting a target. But even people who are considered 'well trained' with pistols have learnt and practiced on a firing range – a one-way range where there is no one shooting back. And someone shooting back at you raises the art of pistol shooting to a whole new level. The sonic crack of bullets passing your ears puts some people off their stroke. At least it can make careful aiming an afterthought. I have just been discussing, with a friend in Germany who trains Special Forces for a living, the idea of having people shooting back on ranges and also taking recruits for Special Forces units specially to hot zones to get them shot at a bit to make them 'steady'. There's nothing like veteran troops who have been shot at a few times for an example of what 'steady' should look like.

The Author teaching dirty pistol tricks to German SF in 2010.

Drag your feet until the gunman gets close and pushes.

Then disable him and take the gun.

Now take the gun off me! This is how you hold a prisoner when you have the weapon. A searcher should approach from the side. (Author's Collection)

My main message about using a pistol when working undercover is that, depending on your situation and mission, very often the speed you can draw the weapon, or how well it is hidden, is as important as how well you shoot. Think about this: suppose three crooks hold you up at pistol point in a South American bar. They don't know you are carrying a pistol so they are overconfident and sloppy. Only one keeps his gun on you and then not all the time. They have almost certainly not been trained on their weapons and every person they have threatened has been afraid of them. You act scared and wait for the moment. You draw and fire into the middle of each body, the most dangerous looking first. Then as they go down you close and put a shot in each head to make sure.

The main idea, I say, is to manipulate the situation so you catch the enemy by surprise. Then you don't have to draw faster than they can pull a trigger. Perhaps you can even get yourself into a position where you have your gun out ready? Then, at a distance of a couple of yards, fire into the body and finish with a shot to the head. The reason for the head shot is quite simple. A lot of people wear body armour which will stop a pistol bullet or reduce its impact. Aim first for the body to be sure of a hit which will stop them killing you and knock them down. Then walk up and finish them off in your own time. There used to be a saying in the British SAS, 'Two in the head and you know that he's dead'. There's nothing like making sure.

## Holsters

Speaking of hiding, drawing and shooting, you will need a holster of some kind to carry your pistol in a 'concealed carry'. As an undercover operator you need to hide your weapon well yet be able to get at it quickly. The balance between hiding and access means there is no clear favourite for holster type. Your holster can go on your same-side hip, against your backbone, in your crutch or under your opposite side arm. You will understand that in a covert situation these holster 'carries' have different advantages and disadvantages relating to speed and accuracy.

Same-side hip under your coat is the fastest for a draw and the weapon is moving up and forwards as it comes onto the aim so it will not be drifting off laterally as you fire. But it is not very well hidden and will be found in a quick pat down. Behind your back may sound strange but it is a good hide when you may get a pat down. If you have worked on checkpoints you must have searched countless people. I bet you checked down each side and never felt their back. The problem, of course, is that it is a little slow to access. In your underpants against your crown jewels is the best place for hiding a pistol because most men will not search another man's crutch. Of course, it is slow to get out and it needs to be a fairly small pistol but then you only need a small pistol. Under your arm in a well-fitting shoulder holster is probably the best for most situations as it is well hidden, quick to access and hard to take off you. The weapon is moving sideways as you draw so it will tend to drift off the target. To counter-act this, from the holster point the weapon down to the floor then up onto the target so it is rising when you fire. Make your decision according to your operational situation.

## Pistol shooting

I am not going to say much about pistol shooting technique here as, although I do teach fancy shooting, this is not what you need on operations. You just need to stay calm and use your common sense. After reading what I have to say about pistol selection I want you to get yourself a pistol and practice, practice, practice. In your barracks get the stripping and assembly fast and sure as this is the best way to familiarize yourself with a weapon and come to trust it. Then practice stoppage drills, with blanks if you can get them, until these don't hold you up any more than a magazine change. Pistol shooting in a combat situation is not target shooting. It is not so-called combat pistol shooting either. It has almost nothing in common with shooting on a range. Your goal, depending on your mission, is to stop someone killing you by shooting at them as quickly as you can or it is to get up close to someone and kill them without a fuss.

My pistol technique is a little different from what is taught almost everywhere but it works for me. Compare what you are taught with the following: draw your pistol with the hand you are going to pull the trigger with. As you come to the aim, grip the base of the hand grip with your free hand, palm facing up, to keep the weapon on the target as you squeeze the trigger. Get the first round off as quick as you can in the general direction of the target then keep shooting until the target goes down. Aim for the centre of the observable mass – body – unless you are very cool, very good and the target is within spitting distance. In which case go for a head shot as there will be no body armour in the way. After your first hit the target is likely to be no immediate threat owing to being shot somewhere that hurts. According to the circumstances you might want to put further rounds into the body from a distance or approach and kill with a head shot. There are many clever techniques employed by pistol experts which lead to greater accuracy and rate of fire but they are more an art than a science and have little place in combat. Your job is to kill your target before he gets you – not put on a circus act.

## Working undercover

The one advantage of a pistol over a rifle, from the SF operator's point of view, is that it can be hidden when you are working undercover. A pistol on show, like a knife or a rifle, simply means no one will mess with you unless they have a weapon or get the drop on you. If they want to kill you they will smile and chat until the time is right. Then you will be dead without getting a shot off. If you carry your pistol hidden, on the other hand, you are in a position, should it suit your purpose, to allow people to show how friendly they are before you draw. A person is likely to threaten an apparently unarmed operator before shooting or stabbing them. And carrying a hidden weapon allows you to be armed where others are not – like a bar.

Undercover operators do get to use their pistols now and again but perhaps they don't carry the sort of pistol you expect and probably they don't use it in the way you think either. A couple of months ago a good friend of mine working in the Balkans was sitting in a bar in a rough part of town. He was meeting someone there and dressed in the typical British expat-abroad outfit – jacket and slacks – but a bit on the scruffy side and he looked pissed. This was where all the business was done in town and he was selling a truckload of Russian cigarettes and hoping to make some vital contacts. The place was full of drug dealers, weapons dealers, the odd Al-Qaeda operator and the Intelligence operatives of more than half a dozen nations keeping an eye on it all and each other. Two big, crop-haired, guys in suits and dark glasses walked in and looked across at my friend. One stood near the door and one walked towards him and made the mistake of reaching as if for a gun. As if they were amateurs coming to make an arrest. My friend drew a tiny .22in automatic from a rear belt holster and, aiming under the table, shot the furthest guy in the knee so he went down screaming like a stuck pig. In a flash he was on his feet with his pistol stuck in the mouth of the nearest suit. The standing suit raised his hands wide eyed and the situation calmed down. As much as it ever got excited – it being that sort of bar.

It turned out that the suits were from a supposedly friendly organization and thought they could act like they would at home. Wrong. They only wanted to talk to my mate and should have approached in a clearly non-threatening manner. But I'm not training Spooks here I am trying to give you some idea of what pistols can do. To tidy the story away, these guys were so unhappy they had a bomb placed in my friend's car which subsequently killed his local driver. My friend was quickly and quietly pulled out of the country to avoid a diplomatic incident between friendly countries. Of course no one has a monopoly of blue-on-blue do they? If you know what I mean...

## Choosing Your Pistol

Many operators carry a .22 automatic as the tiny automatic can be hidden in a trouser waistband or holstered pretty much anywhere else. In use it is relatively quiet and a head shot will kill. On the other hand, if I were going to a fight with pistols I would take the 9mm automatic loaded with armour piercing because it will always knock a man down.

The Walther PP series are all excellent and include .22in and 9mm calibres. The Walther PPK 9mm is small and easy to conceal. The Glock 9mm is good too and both have all manner of useful features. These pistols are relatively small yet the hand-grip is chunky and easy to get a grip on.

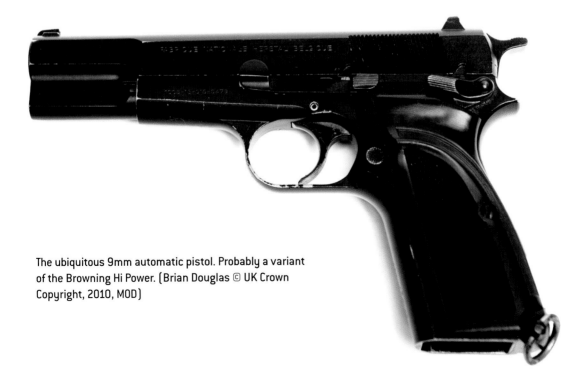

The ubiquitous 9mm automatic pistol. Probably a variant of the Browning Hi Power. (Brian Douglas © UK Crown Copyright, 2010, MOD)

Don't select something like a .40 that fires expensive or unique ammunition or you will not practice. In most theatres, 9mm is the easiest ammo to get hold of and therefore cheap or free. If you get the choice, go for something light if you are going to have to carry it all day.

The Heckler & Koch P7M8 or P7M13 in its 13 round 9mm form is a superb weapon with many outstanding features. There is a system for automatic safety and another to stop you burning your fingers on the hot casing. Being German, the engineering and design are superb but you will have to buy one yourself if you want one because I don't think anyone but the British SAS issue them. For many years I carried a Berretta 9mm as I found the hand grip comfortable, the magazine carries a couple more rounds than the Browning and I got one cheap. If I am going to give you a guide go for a well-made 9mm so it won't jam and you will have lots of bullets when you need them. The first time you might tend to fire a little quickly.

If you plump for a revolver, any well-made .38 Special is a good all round choice being more powerful than the .38 and utterly reliable.

# KNIVES

Keep your knife for cutting your steak. Probably the only time you will use one in anger is in a bar-room brawl. You are even less likely to use a knife to kill an enemy soldier than you are a pistol. The best reason for carrying a Ka-Bar or Bowie knife is to 'look the part'. But I want to make you a better soldier not look like an extra from *Rambo*. Like any other weapon, if you are in a hostile crowd on patrol a knife could be taken and used against you. Easier than most things if it is in a sheath on your hip or webbing. Don't offer the target.

Knives do come in handy as a tool or as a weapon but the best type to carry is a lock-knife. Perhaps with a saw and screwdriver built in. This is a folding knife which locks open to prevent it closing on your fingers. It can be carried discretely in the pocket where it will not be snatched in a crowd and it can do all the things you need in the field like cutting string and sharpening wooden stakes.

## Goal: When is the right time to employ this weapon? What is it good for?

The one time you might want to use a knife against someone is when you are undercover and unable to carry a gun on the grounds of being unable to hide one. A tiny lock-knife with a 2in blade is plenty to kill with if you use it right.

You have no chance against an opponent with a pistol or even a stick if you have just a knife. The use of a knife in combat should be a surprise rather than facing up against an opponent if you can possibly manage to arrange things that way.

Given you have the element of surprise, you can cut the carotid artery with a short knife and cause death within a couple of seconds. Insert the knife in the side of the neck with the blade pointing to the front and cut forwards. This is messy. Given a longer knife, if you approach from behind and put one hand around the head the target will raise both hands to give you a clear shot at his belly, Slide the knife upwards and tear to cause maximum blood loss. A knife over about 9in will easily reach the heart.

The only way to defeat a knife safely, every time, without carrying a weapon yourself is to wear Kevlar gloves and shirt then grab the knife blade. It works surprisingly well. If you could carry a baton, however, it would be a simple matter to break the wrist of an opponent with a knife. I carry a telescopic metal baton.

## Skill: How do you use the weapon to maximum efficiency?

I have only ever used a knife once in military combat and it would have been much less messy to have used my fists or a rock. If you have to carry a fighting knife, choose a knife with a good grip on the handle as blood is very slippery and you don't want to drop it. Speed and

The Author teaching knife fighting.

Surprise is everything when you need to take a knife.

Take the elbow, then the wrist and then turn the knife back on your enemy. (Author's Collection)

reach is important so go for a light knife which is as long as you can carry. The point is for killing so the blade should be slender for better penetration through heavy clothing or webbing.

Don't draw or show your knife unless you are going to use it. Keep it hidden until you are about to strike. Keeping the knife hidden can give you the element of surprise. If you lose that and it comes to a 'face-off' lead with the knife like a sword-fencer if your opponent has a knife and guard the knife with your free hand and body if he doesn't. My main concern with you carrying a knife is that some people are shy of using them. Showing a knife to deter an attacker in a civilian situation can get you killed if you can't or won't use it. This is doubly true in a military situation. A knife enters the human body with the slightest touch and opens up the flesh without any obvious resistance. But there is always lots and lots of blood and the target's guts often drop out – so get used to the idea.

There are three principle cuts to kill with a knife. It takes a long knife to reach up to the heart from the navel. It takes a long thin knife to go through the ribs from the back and to the heart without sticking. It takes only a box-cutter, as they call them in the United States, (Stanley Knife in the UK) to cut the carotid artery or vein with a horizontal, deep, forward-moving stroke on either side of the neck.

After the events of 9/11 I hope you will not underestimate box-cutters – especially when they may be brought out as a surprise. Many years ago in a bar fight a chap caught me on the top of the arm just below the elbow with a box cutter. I didn't feel a thing but the flesh opened up like a pair of lips and I lost the use of the arm for the duration of the fight. The medic had to stitch up the muscle and tendon inside the arm before stitching up the surface. I just have a 3in scar now but against a better opponent, out to kill me, the next hit could have been fatal. Remember: don't drink and fight – it's worse than drinking and driving.

If you are attacked by someone with a knife then distance is all important – as with all contact sports – especially if you don't have a knife on you. Maintaining a distance will ensure that your opponent has to step forward to touch you. This feels awkward initially but it will give you 'thinking time' to dodge or counter the strike. An open coat held like a bull-fighter can also be useful.

Let the strike follow through on its swing as you step out of the way then move in with a jab to the eyes or a blow to the jaw. It's not simple and it won't work every time but then it is better than nothing. If you have a knife this is the one time you can use the edge of the blade – you slice your opponents knife arm at the wrist or forearm as he reaches out to strike you. This way you can maintain maximum distance and avoid injury yourself.

## Types of knives

The Bowie knife and Ka Bar are good examples of a large solid knife which is handy for camping and will kill someone efficiently if you can hide it from them first. A lock-knife with a 6in, or even a 3in blade is just as deadly, just as useful and much easier to hide.

### Summary: What are the important things to remember about knives?

Carry a small lock-knife on operations to use as a tool. It should be light and easy to hide out of sight so it cannot be taken from you in a crowd situation. As a weapon, the knife is always a last resort for a soldier or undercover operator. A weapon to use when there is nothing else. In a face-off you have a far better chance armed with a stout stick but a small knife can be carried when nothing else can. Try not to get into that situation.

# MAPS AND GPS

There is no point in my giving you a map reading lecture here: you should all have a GPS system (if not invest in one). They never break down and cannot be jammed by the enemy.

Actually, for the innocents reading this book, that last sentence is a lie. All tech kit breaks down and most can be disrupted by electronic counter-measures when it suits someone with the ability. Without your GPS and behind enemy lines you had just better be able to navigate with a map and the stars.

Anyone can read a map – rivers and roads and forests are just drawn on it and there is a legend in the corner which explains everything. It is just transferring the information to the ground and back that takes practice. What I

Tom exits the aircraft posing like a film star: when you are all alone and behind enemy lines you really need to know what you are doing with a map. (Photo courtesy Tom Blakey)

mean is reading a map is easy if it has roads and other useful features on it. Some maps just have endless desert, bush or jungle with, and if you are lucky, a few contours or distant mountains. So they aren't much more than a blank sheet. The less information there is on the map the harder it is to find out where you are and where you are going. There used to be a map of a tank training area in Canada which covered thousands of square miles and had only one tree on it in the middle. Then a drunken British tank driver drove over the tree… What I shall do is give you a handful of tips relating to navigation that have come in handy for me.

## Finding out where you are with map and compass

If you don't know where you are exactly and you can see two or more unique landmarks which appear on your map such as mountain peaks or churches then you can easily find your position with only a map and a compass by using the following system which is called a 'Resection'. I will take you through the process quickly. Take as long as you like to learn it.

1. **Magnetic bearing:** Take a bearing on one of the landmarks. You do this by pointing the sights of your compass at the landmark and then aligning the dial of your compass with the north-pointing needle. This gives you a 'Magnetic Bearing' from you to the landmark.

2. **Back bearing:** Add 180° to this bearing – half a circle – and you will have a 'Back Bearing' – the magnetic bearing from the landmark to your position.

3. **Magnetic variation:** Map (Grid) North is slightly different to Magnetic North so all magnetic bearings are slightly different from map bearings. The reason for this oddity is partly the way maps are made and partly that the Magnetic North Pole wanders about slightly. So each year it will be a little different from the north shown on your map according to the map's age. To correct for Magnetic Variation is very simple: on your map it will say that at the given date of printing the Magnetic Variation was X degrees Y Minutes East or West of True North and changing by X Degrees Y Minutes per year. A Minute is 1/60th of a degree.

   Work out how many years since the map was printed. Multiply the yearly change by that figure. Add it to the given figure when the map was printed. This figure is your map's current Magnetic Variation. It will be a few degrees and minutes.

To change a Magnetic Bearing to a Map/Grid Bearing all you have to do is subtract the Magnetic Variation.

4. **Plotting a back bearing:** Turn your Magnetic Back Bearing into a Grid Back Bearing by subtracting the Magnetic Variation. Plot this on the map by using the edge of the compass or a degree measuring device and drawing a line at the correct angle from north running from the landmark towards your estimated position.

   Do the whole thing again with a second landmark. Then again if you have time and another landmark for accuracy.

5. **Resection:** The back bearings you have marked on your map will intersect exactly on your position if you have done the process accurately. So you will know where you are to within a few tens of yards. If you did three and they do not cross on a point then one is wrong. At least one.

## TOP TIP!

*The rhyme I use to remember the direction calculation is:*

Mag to Grid get rid

Grid to Mag Add — for when you are doing it the other way around

## Marching on a bearing

Once you know where you are, and you want to get to another position marked on the map but is out of sight or a long way off, then there is a way to direct your route called 'Marching on a bearing'.

Measure the Grid Bearing from your position to your goal with a protractor or the edge of a compass. Add the Magnetic Variation to turn it into a Magnetic Bearing. See rhyme the box above.

Set the Magnetic Bearing on your compass and then line up the arrows on the compass with the North Seeking Needle. The sights or arrow on the compass will now be pointing to your target.

If you do not have a protractor handy then lay the direction arrow or edge of your compass on the map and along the line of your intended march. Set the dial of your compass so that

North on it points to Grid North on the map. Then add the date-corrected magnetic variation from the map and you are ready to go.

**Fixed landmark:** Sighting along your correctly set compass you should aim to pick a fixed landmark where it is pointing and use that to march towards. This minimizes sideways drift while marching.

**Artificial landmark:** Where there is no visible feature on the landscape, such as some deserts or jungles, you should send a man ahead some hundreds of yards or as far as he is visible, keeping him on the bearing with instructions. You can then march up to him and repeat the procedure as required.

### Grid reference

To tell someone else where you are on a map you need to give them a grid reference. The idea is that all maps you will use are marked off in grids which form squares usually 1,000m on a side. There are two digit numbers along the bottom and up the side of your map applying to these lines which form the squares. By finding the two numbers along the bottom which leads to the left side of the square you are in, then finding the two numbers up the side which lead to the bottom of your square you can describe it accurately as, say, square 3648. This is a four-figure grid reference.

> **REMEMBER:**
> Use this rhyme: along the corridor and up the stairs; bottom number first then side number second.

To be more accurate, you can make a six-figure grid reference by splitting the box you are in into ten equal segments. If you are just short of half way from the left to the right of the same box and just over half way up it then your position would be 364486. This gives your position to within 100m and this is enough for most purposes.

### Judging distance by time marched

While in training you should accustom yourself to judging distance marched or run by the time taken. Time yourself over known distances. It should become possible to estimate your progress quite accurately when you know that on hard, flat ground you cover 4km per hour in full kit. Then you can experiment with different types of ground and so forth. Notice the difference.

You may be surprised how reliable your pace is in measuring distance covered. This is a very useful technique to master so it is worth some effort. If you have been marching through featureless terrain for six hours and then get hit you may need to know where you are quickly so you can call in a chopper to get out the wounded or you out of the shooting match.

If you know you have travelled 24km because you were doing 4km per hour you can quickly mark it off on your map and read off the **grid reference** to call in the casevac or air support. It is OK to do this in 'Clear' rather than code as the enemy will see the LZ anyway and to use code could in fact compromise the code.

### Contouring

In hilly country you use up most of your energy going up and down hills. If you know that your goal is marked by a distant peak or whatever – so you can find it if you stray off track – then it is sometimes possible to save a great deal of energy by travelling around the intervening hills maintaining the same altitude but going off the direct route. This is called contouring.

### Don't mark your maps

Despite the above techniques for resection etc, we were always taught not to mark maps in case they fell into enemy hands. I always thought this was a little depressing but you can see the advantage of keeping the enemy in the dark. Consider this when I mention capture of enemy maps later. As a compromise I think it best to mark as little as possible on a map that is going out on patrol. Never mark static features such as temporary base camps.

## Finding direction without a compass

**Finding north using the North Star:** North of the equator all the stars appear to revolve around one point, which is exactly due north and marked by a star called, oddly enough, the North Star or Polaris. This star is always directly north of where you are standing. From this exact north marker you can work out any direction you need for movement.

**Finding the North Star:** There is a certain group of stars shaped like a frying pan which also has a proper name. It is called the Plough. Just so you don't think I know nothing these patterns are called Asterisms. See the illustration of the Plough below.

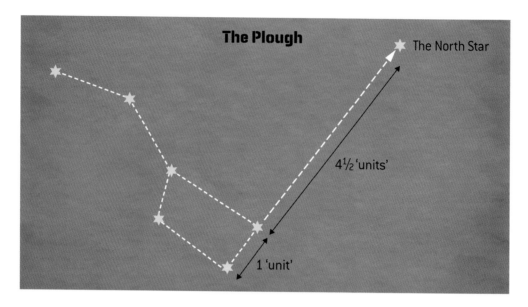

The Plough

The North Star

4½ 'units'

1 'unit'

Even though it revolves around the North Star, the end two stars in the Plough always point directly towards the North Star and remain around 4½ times their distance apart from it. There is one problem with using the North Star for direction: when it is daylight or cloudy you can't see it.

**South of the Equator:** Here you cannot see the North Star and there is no such thing as a 'South Star'. Fortunately there is a cluster of stars called Sigma Octantis, which circle the South Point. If you cannot see these, and they are pretty faint, there is another group of stares called the Crux. Remember: that all these stars circle the South Pole position (where the dotted lines meet in the diagram on p.109) every night just like the Northern group. So you might be looking for them upside down.

**Finding north using the sun:** North or south of the equator you can use the sun to find direction. In the Northern Hemisphere, if you have a wristwatch with fingers, point the hour hand towards the sun. South is halfway between the hour hand and '12' on the dial. North is exactly the other way. For the Southern Hemisphere reverse this entire procedure. This technique is not very accurate but it can be *much* better than nothing if you are stuck.

**Winds:** You may find yourself in a place where the wind always blows from a certain direction. Or even to the shore in the morning or from the shore in the evening. Make sure you know.

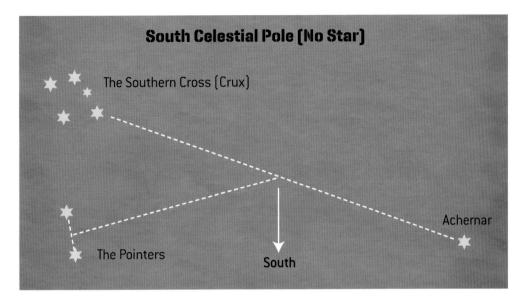

**Rivers and slopes:** In some places all the rivers run in a particular direction e.g. east to west. In some places the ground for countless square kilometres falls to a particular direction e.g. east or west etc. If you operate in such a place it is your job to know these facts.

**Moss:** You may have heard that you can find east by checking which side of a tree that the moss grows on. I have taken a look at this and found that different types of moss grow on different sides of a tree in different places according to local conditions. So it's bullshit. Do not rely on this.

**City lights at night:** If you are lost but know there is a large city within 20 to 50km you could wait for nightfall. City lights can sometimes be seen from an amazing distance as glows on the horizon. If you can see which way and roughly how far a known city is and you know which way is north then you can work out your position by a mental back bearing on the city. Try this to get your head around it.

**Other aids to direction:** If you are stuck out in the wilderness with no aids to direction – even though you know all the above – then the way to find civilization, or at least a village, is to walk downhill until you find a river then walk down the river until you find a bridge. Follow the road which crosses the bridge downhill until you come to buildings or signs of life.

This technique is only for use when you don't know which way to go and cannot see any other signs such as tracks or signposts. The logic behind it is that rivers always run in valleys so you will come to one if you go downhill. Rivers flow downhill and get bigger. People tend to live in lowlands and valleys rather than on hilltops.

### The Global Positioning System (GPS)

This wonderful system tells you where you are at the touch of a button to within a few yards. In case you didn't know it measures the time radio waves take to travel from a selection of satellites. More advanced models can give you bearing to march on, your average speed and make tea. How I wish we had GPS when I was younger.

So the GPS means everything above is a waste of time and effort does it? Did you know there are ways of blocking a GPS signal? Did you know there are machines which will send out an electromagnetic burst and knock them out together with all your electronic kit? This is why the latest aircraft are fitted with girocompasses. And of course all gadgets break down.

### Summary

Practice your map reading because otherwise you are in it up to your neck if your GPS breaks down. Gadgets always break down just when you are relying on them. Don't mark maps with anything of value to the enemy – they may be found or captured. At the very least, learn everything I just told you.

# NIGHT VISION EQUIPMENT

## The benefits of night vision equipment

At night it is typically too dark for human eyes to see who you want to shoot at or, if you can see their outline, you can't see down your sights to take good aim. It doesn't take a great deal of military wisdom to realize that, in the dark, anyone who can see is going to have a 'no contest' win when it comes to a shooting match.

Just imagine, you are out on/in some dark mountain/moor/forest/jungle/suburb and the enemy is out there too. You just don't know where the enemy is and they don't know where

My mate Sadie testing the first ever infra-red night sight in the British Army, Yemen, Radfan, 1965. I'm pleased to say that the technology has improved considerably since those days. (Photo courtesy Colour Sgt Trevor 'Sadie' Sadler, 1st Battalion 'The Vikings' Royal Anglian Regiment)

you are either. All warfare was like this when I was a boy. If you have the ability to see in the dark then if you come across the opposition you can shoot and hit them while they are blasting off anywhere. If you set an ambush it is like shooting fish in a barrel. If they have night vision and you don't then just don't play at night. It is suicide.

Over the last 35 odd years, since I first used early passive night vision equipment (the 'Starlight Scope') in Northern Ireland, technology relating to this marvellous kit has come a long way. The early night sights gave a green picture and switched off from overload when you pointed them near a streetlight. They were also big and heavy – strapped to the top of your rifle making it both awkward and seriously unbalanced. At one point the night sights weighed more than the heavy rifles we carried.

Today, if you have the right gear, you can see anything, anywhere, no matter how dark the night. And the equipment has shrunk in both weight and size beyond all recognition so you can wear it as glasses and see in total darkness. Night vision equipment is issued to pilots, drivers and it is, or ought to be, issued to all infantry too. By the time you read this the technology will have undoubtedly moved on yet further so I am going to explain the principles of night vision equipment and how you use it properly rather than specific models.

## Normal vision

Our vision works by our eyes receiving light which has reflected off the object we are looking at and onto our eyes but which came originally from a light source such as the sun, moon, stars or a torch. In the 'dark' our eyes cannot gather enough of this reflected light to make out the object.

## Night vision

There are five main ways of fixing a shortage of light so we can see in the dark:

- You can shine a light onto the object to generate more reflected light
- You can shine a wavelength of light onto an object which is invisible to the naked eye then view it through a gadget
- You can view the thermal radiation coming off the object as if it were visible light
- You can gather more available light coming from the object and get it to your eyes
- You can gather the light coming from the object and make it stronger electronically

**1. The Torch:** The simplest way to see something in the dark is to shine a torch or searchlight onto it. The obvious disadvantage is that the enemy can see your torch and therefore where you are. Close enough to be uncomfortable at any rate. An alternative is to use flares of various kinds but then the enemy know they have been seen and flare light is neither clear nor long lasting.

**2. Active night vision equipment:** Active night vision equipment projects an electromagnetic beam like a torch onto the target and then picks up the reflection and turns it into a picture the human eye can see. This beam is normally infra-red which is just below the frequency visible to the human eye but it can be a type of laser light. It is like shining a torch on the target that no one else can see. The problem with active night vision equipment is that anyone else with a receiver for, say, infra-red can see where you are looking from as plain as if you were shining a torch. This kit used to be issued to drivers at one stage so they could see their own infra-red headlights but the practice pretty much stopped when both sides had the kit.

**3. Thermal imaging equipment:** All objects warmer than -273° give off heat for complicated reasons to do with black-body-radiation and quantum. This heat can be picked up and turned into a picture by thermal imaging equipment because heat such as you feel from a fire is actually an electromagnetic wave similar to light but at a lower wavelength. This type of kit is excellent because it does not give you away and can see through light fog, rain and smoke – but at the time of writing the picture could be better.

**4. Night glasses:** These are pretty ordinary binoculars except the object lens – the one furthest from your eyes which collects the light – is much bigger than normal. But just like a normal pair of 'binos' it focuses the light it collects onto the rear lens and then onto your eyes. Because of the larger size of the object lens more light is collected than by your eye and therefore the image is brighter. Quite a lot brighter. Until the advent of electronic night vision equipment this was all we had and night glasses were widely used particularly at sea where object lenses of over 2in were used. There were also some special eye-drop drugs developed to use with night glasses which made your pupils open wider to collect more light – so count yourself lucky that technology has moved on.

**5. Passive night vision equipment:** This is the kit you want to use if you can get it. Essentially it is a TV camera which collects whatever light is available coming from the target. And there is always a tiny bit of light however dark it seems to the human eye. Then the mechanism turns up the brightness with an amplifier and shows the result on a screen. Through the wonders of science this can all be made so small you don't realize how clever it is. Helmet mounted you hardly know you are wearing it. It does need a battery power source but it does not give you away and can see clear as day in the darkest night.

## Human night vision

The human eye has two types of light receptors at the back of the eyeball. They are called rods and cones because that is what they look like. Small ones obviously. The cones see colours well but not in bad light and the rods see black and white best and they work fairly well in bad light. The rods taking over is why things lose their colour and look various shades of grey to us at night.

The cones are placed mostly in the centre of the receptive patch in your eyeball and the rods mostly around the edges. This means that you see things in colour that you look straight at

Night-vision goggles that can be mounted onto helmets. (USMC)

but not so well out of the corner of your eye. This is not obvious because your brain fills in a lot of the picture from memory. What matters to you is that your best night vision is obtained by looking at a target with your peripheral vision – out of the corner of your eye. Try this at home as most people have to test it to believe it. If you have a dog you may have noticed that he or she can see really well in the dark. This is because dogs have a lot more rods than we do. Dogs, cats and many other nocturnal animals also have something called a tapetum lucidum which is a membrane at the back of the eye. This sort of reflects more of the light coming into the eye to where it can be sensed by the retina and this is why you can see cats' eyes shining at night.

Another feature of the human eye is that it takes it about 40 minutes to get used to bad light. You must have experienced coming from a light place into the dark – driving into a tunnel say – and being blind for a few moments. Well for some reason – and I don't know exactly what it is – it takes your eyes about 40 minutes to reach their best in bad light. The thing you need to remember is that a moment's exposure to a bright light, when your night

## TOP TIP!

### Always use passive night vision equipment

When passive sights were in an early stage of development they were not that good if it was very dark and active sights were much better in that situation so they had a use. Now that passive equipment is so good it can work efficiently no matter how dark it is. There is no obvious benefit to active equipment and a fairly serious drawback – it could get you shot.

vision is working, ruins it. And then you have to start the 40 minutes all over again. While you are doing this you are vulnerable so guard your night vision. If you have to fire your weapon or take a look at a map by torch light then close one eye – the one you shoot with if you are not shooting at the same time. This protects the night vision in your aiming eye for that all-important first shot. Muzzle flash ruins your night vision instantly so you should aim to keep your spare eye closed while you are in a shooting match.

## Summary

Always carry passive night sights – one per man if you can get them. Never use active night sights as they will get you killed. Use night vision equipment to read maps etc. at night rather than a torch because this will preserve your team's 'night vision'. If you have to use a light, warn the other members of your team and close one eye yourself so as to maintain night vision.

# RADIOS AND MOBILE PHONES

## Goals: Why communications are important

One of the reasons an army is always superior in performance to an armed mob is that confrontations at every level are won by bringing superior force to bear against the enemy at a certain point. All armies are organized in the hope of achieving this if they can. This strategy is called achieving a localized superiority of firepower – and is covered in some depth later. To bring a superior force to bear requires someone to make decisions and issue orders, people prepared to obey these orders and, of course, that these instructions be received. No communications means a headless mob not an army.

What allows an army to stay organized is its communications. It is necessary for commanders to be able to tell their troops where to go and what to do. It is even better if the troops can report back what is happening so the commanders can assess the success of their plans and send reinforcements or make other adjustments as appropriate. Without effective long-range communications no large-scale strategy can be put into action. How could a commander control the movements of a division or a battle group without being able to tell his men what to do?

## History: Forms of communication

For thousands of years the signal to charge was given by blowing on a conch shell or a trumpet. It was something but think how limited this was in terms of what a commander could tell units at the other side of a battlefield. Romantic as it is there are limits to the detail which can be conveyed by a bugle call. Though there were a number of bugle calls which every soldier had to know: advance and run away being quite important amongst these. Bugle calls were messages from a leader to the whole of his command unless he had been able to issue orders to the effect that, 'When you hear the bugle just the left flank move forward.' What about orders for the artillery to move position or the cavalry to take the left flank or anything in reaction to what the enemy is doing?

This is where the messenger came in. It took time to tell or write the message, time to get it there even if the messenger wasn't intercepted and time to get a reply back if the officer on the spot could see a better alternative or had a report to make. The Charge of the Light Brigade (1854) during the Crimean War was deeply heroic but resulted in the total loss of an entire British Light Cavalry Brigade! This was all down to bad communications. The Commanding Officer, Lord Raglan, issued the order for the brigade to attack the guns to their

Vietnamese bugler sounds the charge: Vietnam, 20 January 1966. (Corbis)

front – meaning a certain small battery of artillery. The order was misinterpreted for various reasons, including petty bickering amongst the officers, and the brigade, led by its senior officers, charged the wrong guns.

They rode down a valley with rows of guns on either side and to their front. They were cut to pieces – though some men got to the other end and destroyed a few guns – but the Russians were shocked at the stupidity. Remarkably, the British cavalry knew the order was in error before they set off but such was the discipline of the British Army that they carried it out without question. This whole disaster was down to bad communications and garbled instructions. Once the charge began there was no calling them back because they didn't have the ability to send signals faster than a horse could gallop. Remember this when you are relaying instructions by mouth or on the radio.

Before radios the greater part of a battle had to be planned in advance and the control an area commander had over his regional/area/group commanders was limited by the speed a horse could travel. This is why the armies were laid out for what was called a 'set piece battle' – sort of like chess.

## The use of radio

The primary form of communications an army has today, both at tactical and strategic levels, is radio. Radios at all levels of command have been around for so long now that there are no serving soldiers who have had to work totally without them. But instead of there being one radio per platoon as not so long ago, there is now radio at section and even individual level. With radio, two-way voice communication is effectively instantaneous. Orders can be passed down from any level to any level below and reports can travel the other way at the speed of light. This is a tremendous advance on messengers if the orders are good and the people at the other end understand what is required.

**Voice procedure (VP):** Radios don't make up for bad orders but in an effort to achieve clarity and avoid misunderstanding everyone using a radio is taught a type of voice procedure which varies slightly between armies but amounts to the following:

Each radio set has a callsign which is known to the other callsigns. These are often arranged in a tree formation so within a battalion radio net there would be callsigns for each squadron, troop, section and so on down to the individual radio set. An example might make this plain: C41 might be C for C Squadron, 4 for 4th Troop, 1 for the first section or fire team. Then there are specific titles for motor transport, mortar group and so on to identify every fire group and every significant sub-unit in a battalion.

Once all this, and quite a lot more, has been committed to memory it is simple to call anyone in the battalion without someone else thinking you are talking to them. This is how it is done: 'Hello C41, this is C44 is Sunray C at your location over.'

So here you have stated who you are calling, who you are and asked if Sunray, the commander of C Squadron, is present at that location. Using 'over' at the end of a message means you have finished speaking and expect a reply.

Then there are 'radio checks' where the lead callsign checks the reception of each of his units and much more. I can't give you a radio course here, and in any event you need to learn the VP for your own army, but you will get the idea.

**The problems radio brings:** Having the use of radio communications does allow marvellous communications amongst a unit and solves many problems for today's soldier but it also brings one or two problems unique to itself:

- **Too much talk:** Of course, with everyone being able to talk, and everyone on the same frequency being able to hear, there is lots of room for confusion and one person talking can block another person with an important message as only one message can be sent on one net or frequency at a time. To keep the volume of messages down to a manageable level for each military unit the radio net is generally split up along lateral bands: there will be one frequency, say, for platoon A to speak on amongst themselves, another for platoon B and another frequency for the platoon commanders to contact their superiors. This is achieved in slightly different ways with armies around the world but it amounts to limiting the number of people talking on each frequency.

- **Position marker:** Using your radio, or even having it switched on, can give away your position to an enemy with relatively simple direction finding equipment. Two such outfits pointing to your set give the exact position of your transmitter. To avoid this, the French Resistance in Occupied France during World War II for instance, would encode their message before switching on their set in a new place every day. Then they would read off the pre-prepared message as quickly as they could and switch off again before moving.

- **Radio messages can be intercepted:** Bear in mind that every message you send might have someone unfriendly listening in. It could be a member of a terrorist group in a hut in the next street listening to everything you say, or manning a nearby radio listening post or it could be the intelligence agency of a foreign power. There are interception services run by many nations around the world. In some circumstances they might be listening to you. Even neutral, or officially neutral, country might pass on useful information to your enemy.

  The significance of eavesdropping on your communications depends to a great extent upon the level at which the communication has significance and the time over which the communication will be of use to the enemy. As a general rule a message between the leaders of large units is worth more effort to the enemy to intercept owing to the size of the units involved and potentially compromised. The time over which the information remains useful is likely to be greater too owing to the time required to prepare and move such units.

Side view of United States Air Force E-3A aircraft flying with raydome mounted on top and Airborne Warning and Control System (AWACS). (Andy Holmes © UK Crown Copyright, 2002, MOD)

The leader of an enemy or neutral nation is not going to be interested in the details of a corporal's flanking attack on a bunker but he may be very interested to hear that a couple of divisions are being moved from one country to another. Short-term instructions between elements of small units cannot generally be intercepted and acted on by the enemy owing to lack of time unless the situation is very unusual. An exception might be the order to advance on a village in a counter-insurgency operation where the enemy have a local listening post.

There are two things which work in favour of radio security. The first is that very often our enemy speaks a different language to us so they cannot easily understand what we are saying. This means that very often the interception has to be run past a more senior figure in their organization and this delays the passage of useful information between them. This cuts both ways though.

The second is that we can use codes. Simple things like allotting random names to positions and units can make it quite difficult for an outsider, who may not use the same language, to make sense of what is going on. Of course, you will go to more trouble encoding more important information and information which has a longer useful life.

■ **Radio jamming:** It is quite simple, in certain circumstances, for one side in a conflict to block everyone's radio communications but this is not often done. The reason is that to maintain a state where no one can use a radio – generally by a type of broadband transmission – requires a powerful transmitter on the ground, at sea or in the air. This transmitter is a prime and easy target for the enemy and so you must control the area where it operates. If you control the area – land, sea or sky – then you are the dominant power and will generally suffer more from the obstruction of communications than the enemy does.

■ **Radio tips:** There is a lot to know about radio and I can only give you an outline here to get you thinking. There are, however, one or two general rules you can follow in order to preserve security: don't use your mates' real names as the enemy might start talking about/to them and damage morale. They might even track your family back home! Don't use code designations where the enemy can see what you are referring to as this compromises the code. This means don't say, 'We are approaching Zebra now' as you come back to base camp. This gives away the code name so that whenever 'Zebra' is used in future the intent may be clear. Keep your use of the radio short as it can be tracked. There are even weapons which can home in on your transmission. Use the radio then move quickly especially when you are a small unit and the enemy may have the ability to find your position and come visiting or drop something nasty on you.

## Mobile telephones

Recently soldiers have started to use their own mobile telephones in barracks and on the battlefield too. Very much the same security principles apply to mobile telephones as radios. They are simple to listen-in to from anywhere in the world — for a country with the right equipment. Someone with the right gear listening to you chatting from Afghanistan to your wife or girlfriend back home in the States or the United Kingdom can get her number. And from that her exact position within a few metres. Do you really want that?

Because mobile telephones use a digital system they are even easier to monitor by machine than a radio transmission. There are now machines which listen to conversations and pick up certain words such as 'train', 'plane' or 'bomb' — as the most obvious examples. These conversations are then automatically recorded, tracked and flagged for closer attention by humans. That is great for catching amateur terrorists but there are countries around the world interested in other things you might be saying.

While a radio signal gets distorted over distance, to all intents and purposes, a mobile telephone doesn't. Because the digital mobile telephone signal is bounced from the handset to a local booster and then up to a satellite, off to another satellite and

Satellite radio in use at a FOB in Afghanistan.
(Fiona Stapley © UK Crown Copyright, 2006, MOD)

then back down to a booster and finally to the receiving handset it is a relatively simple to monitor any telephone conversation from anywhere in the world.

Most intelligence is not 'James Bond' stuff at all – it is low-level information aggregated up. Without giving too much away, the radios in a large unit would be used more before an operation wouldn't they? Likewise, mobile telephones owned by members of a unit might all be used to book trains or planes before going overseas. This could tell a foreign power more than you might want them to know about your troop movements. Watch what you say and listen to instructions from your officers.

## Battlefield internet

You probably already use the internet yourself and may even know that it was invented by the British but set up properly in the US as a decentralized communications system which could not be easily compromised in a nuclear war. What you may not know is that a lot of high-end kit from aircraft to artillery are now using a sort of military internet to digitally transfer maps, pictures, targets and everything else you can think of which might be useful in a battle. Of course it is instant but also the accuracy of the written word and the picture gives a massive advantage over radio. On top of this, the latest machines can even talk to each other. A target spotting aircraft can send the coordinates of a target direct to our artillery or helicopters. And there are artillery pieces which can lay on from this information and shoot all by themselves. Clever hey?

## Summary

My position on communications is that high-tech kit is great. It lets us take down the opposition with less risk to ourselves and generally moves the odds of survival in our favour. On the downside, all gadgets break down and the enemy are likely to help them break down if they possibly can. Set up all your operations taking into account that communications may go down at any time just like that bloody GPS.

Remember not to say anything useful to the enemy on air. Never use real names or titles on air. Follow any and all instructions for the use of code as one slip can give everything away. Keep each radio transmission as brief as possible. When there is risk of enemy fire, move after each transmission. Ensure that what you say is correctly understood.

# CHAPTER 3

# Food, Shelter & Dealing with the Weather

You are only any use to yourself and your mates for as long you are fed, watered and protected from the weather. When you are wet, cold, short of sleep, hungry or thirsty, you are a liability to yourself and your comrades because you are more interested in getting comfortable than getting to grips with the enemy. You might be able to point a rifle when you are chilled to the bone but you sure as hell aren't going to think straight or keep a reliable watch for the enemy. Remember: any fool can be cold and hungry but a good soldier can make himself comfortable wherever he is placed.

It should be your major priority that you make sure you are well fed and watered, as rested as you can be and as warm/cool and dry as is reasonable given the situation. As they used to say in the British Army: eat when you can, sleep when you can, shit when you can.

For you to be able to achieve this you need to be able to do two things: make a shelter from the sun or rain, depending, and find then prepare water and food. Learn to do this efficiently and you will not only make a better soldier but you will make every posting and mission a lot more comfortable for yourself. You might even enjoy yourself; getting paid to visit interesting, exotic places and shoot the locals.

# DRINKING WATER

Like air, water only becomes an issue when it is in short supply. As a rule, this happens in warm countries where there is little surface water, no snow and few taps or water faucets. When you are travelling in vehicles, you can carry all the water you need. When you are in an organized camp or position then the officers should have plenty of water supplied or they are not much good.

Water supply gets more interesting when you are doing proper soldiering; working as infantry on foot and away from resupply for a number of days. In some places you have to carry all your water on your back and water

First wash in a month after an extended patrol somewhere in Africa. One of the men scrounges water from an armoured unit. Armour always have plenty of water – infantry don't when they are travelling on foot. (Author's Collection)

gets surprisingly heavy. Every litre weighs exactly one kilogram but feels more. You also come to notice how much you drink.

In hot countries, especially, make sure you carry sufficient water for the expected length of your patrol or the period until the next resupply so you can avoid sourcing the water locally if possible. Remember without water you will cease to be effective in a day or two and die just a couple of days after that.

## In camp

In hot climates such as North Africa or the Middle East you will require a surprisingly large amount of water to balance what you lose through sweating. In camp you are likely to be able to drink your fill whenever you like so it is not an issue. What does become an issue though, when you are drinking like a fish, is the huge amount of salt you lose through the extra sweating. This salt has to be replaced from somewhere and in a hot climate, where you are drinking your fill, you may be sweating so much that your food does not supply what you need. Often in a hot, dry climate you do not realize how much you are sweating because it dries instantly. You may notice a crusty white covering on your skin – this is the salt you are losing: taste it.

Take advice from your medical staff as to the amount of water you need to drink and ***do remember*** to take the salt pills with which you will be issued to replace your losses. If the salt in your body – the medics call it electrolyte – reduces too much you will get muscle cramps, nausea, vomiting, dizziness and then become very, very ill. Take your salt pills at the same time as you take your multi-vitamins so you remember.

## Water discipline

Despite what I have just said above, I am talking to soldiers not little boys. The fact your body feels thirsty and the sun is shining does not mean that you need to drink all you feel like drinking while you are out on patrol. There is a limit to how much water you can carry and that limit fixes how long you can remain operational – or stay on your feet. By exercising a little self-discipline you can extend this period dramatically from a day to maybe four days without carrying so much water that you can carry little else. I have always carried two 1-litre water bottles on my belt and ammunition-filled magazines in 'chest webbing' in case I lose my back-pack. In my back-pack I carry two further water bottles, at least, and more ammunition plus explosives, comms gear, sleeping gear etc.

On long-range reconnaissance patrols behind enemy lines, such as this SASR patrol in Iraq, conservation of ammunition and food/water supplies is crucial. (ADF)

On extended patrols without support I have always taken just a couple of sips during the day as the water comes straight out through your skin if you drink more. It just takes a little self-discipline to wait for the evening and make a nice brew of tea or coffee; then you keep the water in as you rest in the cool of the evening. This way you should be able to survive and operate without any discomfort on a couple of litres a day whatever the health freaks say. If push comes to shove you can extend this to four days by limiting yourself to one litre a day. I'm not saying it's fun though.

Your officers will plan patrols in such a way that you are re-supplied with water or return before you run out. Clearly, on foot you cannot carry the water to last for more than a few days

## TOP TIP!

### Sourcing water from thin air...

If there is no water available from streams or standing water then condensing it from the air with plastic sheets is possible but is very time consuming so it is really only useful in truly desperate situations.

whatever you do so in areas where you cannot re-supply in the field this limits the effective length of patrols. Of course, in war plans go wrong. All the time. My advice is to carry more water than you think you will need. If you are injured or pinned down by the enemy it is the one thing which you will really need – even more than ammunition. You are just as dead from thirst as from an enemy bullet.

Maybe I'm being a little over-cautious and old-fashioned here. I've regularly had to live off the land while on foot patrol in Africa and elsewhere for a month or more at a time without re-supply. And I've come to no harm so I do know a little about getting by. I have at times carried fruit-flavoured sugar crystals to turn water into fruit drink as a treat and to cover the taste of stinking water. I think it is good for morale to have a little 'treat'. Watch out for soldiers who add alcohol to their water. Boredom and stress can make guys do crazy things. They need taking aside for a quiet beating as they are a liability to themselves and their mates.

The modern soldier spends a lot of time riding in a vehicle – in which case there is all the water you want – you could even maybe wash or clean your teeth. And if you are on foot then you can very often get a helicopter re-supply. But you never know what is going to happen – or go wrong – so be prepared and always keep some water by you. You don't want to survive an ambush that wrecks your vehicle and radio and kills your mates only to die of some disease with only a Latin name which you caught from the contaminated water full of camel shit you drank in the next village.

## Sourcing water in the field

So, you are out on foot patrol enjoying the fresh air and sunshine and you find you have empty water bottles you want to top up. Where do you look? Well you don't do a Ray Mears and start digging up cactus roots for a start. Why? Because Ray is the equivalent of a professor of survival and can find water where the lizards are thirsty and food where the rats are hungry. You are not. And much, much more importantly, Ray has all the time in the world to find food and water with no one shooting at him.

There are certain roots you can extract water from and you can drink urine and animal or fish blood but this is not principally a book about surviving against the elements so if this is your interest you might get yourself a book from an expert like Lofty Wiseman (legendary SAS Sergeant Major and my boss when I was a youngster) on surviving in the specific area you are operating in. It is also Sod's Law that you will find little or no water without going miles out of your patrol route and that which there is may be contaminated or poisoned against use by Western troops. Or camels may have drunk from it – and we all know what disease camels carry don't we?

The way **you** will find water is to find people. And ask them. All people the world over need to drink regularly and set their lives up accordingly. The locals **will** know where the water is for sure. Although in many third world countries the women still have to walk miles each day to fetch the family's water in a tin can carried on their head so don't expect it to be close.

It is up to your discretion how firmly you ask for water. Absolutely always make sure the locals drink first. However much they smile while you have a gun in

Soldier with a CamelBak, a handy piece of modern kit.
(Mike Weston © UK Crown Copyright, 2005, MOD)

your hand, the locals in many areas you are operating in will hate your guts and will kill you given half a chance.

Then wait for a while and then use anti-bug tablets (see below) on the water. Although the water may still look yellow and muddy, and smell like it has just squirted out of someone with dysentery, when the pills have done their mojo it will do you no harm. If you think you would never drink water like this, then you have never been hungry or thirsty for a few days – your pickiness diminishes no end after a while in the sun with no water.

The locals in many places are amazingly tough and tolerant of contamination which would kill Westerners or at the best leave us stuck in the latrine for days. So be careful what you drink. Where possible try to eat the local fruit such as melons to top up your supply of liquid. Even where the water is poisoned or contaminated the fruit will be ok – providing it hasn't been tampered with.

## Water sterilization tablets

On a long patrol you might have to drink some pretty questionable water. Boiling water, contrary to popular myth, does not always kill all the bugs in it. It certainly doesn't get rid of poison. Besides rat poison additives – warfarin – water can carry germs for a thousand diseases including cholera, worms and other creatures which will make a home in your belly and eat you from the inside. There is a bug in West Africa which turns your insides to a bloody soup that runs out through all your holes; and it kills everyone that gets it in a couple of days.

The water sterilization tablets will kills all the germs, parasites and bugs in the water so, when you have assured yourself as best you can that the water is not poisoned fill a bottle and add the sterilization tablets in the recommended amounts. If you are not in a hurry, draw straws for someone to try the water first and wait a few hours for results or the lack thereof.

# FOOD FROM THE QUARTERMASTER

It's traditional to complain about army food. You are not the first and will certainly not be the last. Soldiers have complained for thousands of years and they are not about to stop any time soon. The fact is, your mother has spoiled you rotten and whatever the army chefs cook

British Army field kitchen – hot food is always good for morale. Unfortunately you are unlikely to enjoy such luxuries when on extended SF patrols behind enemy lines. (Photo courtesy Sergeant Roy Mobsby)

up you are going to complain. And if you are pissed off with being away from your girl then the food is as good a thing as anything else to complain about.

But remember: while you are in camp you are going to be waited on hand and foot with nourishing, cooked food supplied hot and ready to eat in huge quantities by happy smiling cooks. OK, maybe the cooks are mouthy sods with a bad attitude but you get the idea.

## Why do you need to eat?

This sounds a pretty stupid question but you may not have thought about the answer. We use food to generate heat and keep our body at the right temperature, to give us energy and to repair damage to our body.

There are four main food types and they are useful in different ways from a soldier's point of view. All these food types can be broken down by your body to give you energy for heat and movement. The amount of energy available from each food can be measured and expressed as calories per weight. Some foods have far more calories than others for the same amount and are therefore compact sources of energy. Chocolate is at the top with breads and similar a little further down the list. If you are an expert then you will realize I am putting this simply:

**Sugars:** All food is broken down to a simple sugar before it is burnt as energy. If you eat sugary food then this will break down quickly and easily to give you a quick burst of energy or recover your losses to some extent. Besides chocolate bars of many kinds, you can buy bars of solid sugar flavoured with mint as an energy boost and some units supply sugar pills — which we used to call K pills from the fanciful idea that each pill kept you going another kilometre.

**Carbohydrates:** Bread, biscuits, pasta etc are rich in carbs and these are your main source of energy. They break down slowly in your body and release energy to keep you going for the long haul. If you have a long march to do eat plenty of carbs the night before. Of course, if you eat too many carbs they don't get used and are then stored as fat.

**Protein:** Meat is the most convenient source of protein. Contrary to what most people think, protein is most important for growth and repair of the body rather than energy so adults don't need a great deal. Aside from the small amount we use to repair our muscles after exercise, the remainder is broken down to provide energy in a similar way to carbs but with slightly more difficulty for the body.

**Fats:** Despite the common belief that fats make you fat, they don't – at least no more than carbs do. Fats such as cooking oil and the fat in meat and other dishes are packed full of calories which will break down slowly and give you long-term energy for action or heat. Eskimos eat massive amounts of fat from the animals they hunt and their bodies use it to keep warm in the Arctic. And you don't often see a fat Eskimo.

## Keeping warm

In cold climates we use a great deal of our food to keep our bodies at their operating temperature. This is why we generally feel hungry and fancy heavy, carb-rich food like stew in the winter. If it is cold where you are working then you will need more food than when it is warm weather just to keep yourself warm.

## Energy

The way the body breaks down food and turns it to energy is far too complicated to explain here. Put simply, you have a certain amount of ready-use energy fuel in your muscles and blood. You use this when you exercise or do a march. When you are tired this supply is used up and the waste product, lactic acid, is held in your muscles and will later lead to stiffness. This is why you should warm down after exercise to flush out the lactic acid with fresh blood.

After a long period of exercise which has depleted your ready-use fuel reserves it takes your body some hours to replenish them and this is why you need a rest for the best part of a day before you are ready to perform again to the same standard in a march or other exercise.

The Author, left, grabbing a sandwich and catching up with the lads back at base camp. Refuel with food whenever you get the opportunity. (Author's Collection)

Some people's bodies replenish the ready-use fuel more quickly than others and therefore they recover more quickly. This is the reason, on a long patrol or series of selection marches, why some people who may be fit and athletic gradually lose condition day by day – their body is not resupplying fuel quickly enough to keep up. This principle is also why some people have more stamina than others.

## On patrol

But what happens when you get out on patrol for a few days? You get to eat ration packs don't you? Armies all over the Western World have derisory names for ration packs but the fact is they are packed full of vitamins, calories and goodness so as to keep you on your feet and maybe even smiling.

Ration packs come in two forms: a sort of bulk pack with which someone designated cook can feed a team of 10 men with and single-man packs which are issued individually. As a general rule, the packs which are meant to feed one man for one day on operations can easily be made to last two days. In the British Army these packs contain a great many tea bags – other armies have coffee too. Of course, with the packs comes loo roll, tooth powder, chewing gum and all the comforts of home.

## Food intake

Food works the opposite way round to water with regard to climate: the colder it gets the more food you need. In very cold climates you need thousands of calories a day to keep your body system working properly and keep warm. In temperate climates you won't take much hurt from not eating for four days but you might want to eat more regularly than that for comfort and to maintain your energy. Four days without food will not injure or impede you overly but after that you will begin to weaken noticeably.

## Hot food

The only thing you have to do with ration packs is warm them up – if you have time. It used to be the case that you were supplied with a little solid fuel stove called a Hexamine cooker from the fuel pellet which provided the heat. You open the tin frame and take out the box of fuel pellets then place one back in the frame which forms a stand for the cooking vessel. One pellet would boil a litre of water in normal conditions though a strong breeze made them a lot less efficient.

Modern rations are even more convenient – many have self-heating packs where you just pull the strip and the pack of food warms itself for you. Wonders of science hey?

But the thing about hot drinks and hot food is that they are mainly for morale and occasionally to warm your body. The food in ration packs does not need to be heated to cook it as that has been done long ago. Why does cooking matter? Because to get the goodness out of many foods the human body needs them to be cooked as this breaks down the cells with the goodness in them. More about that in a moment. For now, be aware that all the food in your ration pack can be eaten cold if necessary and you will still get the energy from it.

## Energy drinks

It was a habit of mine years ago to more or less live on black tea with lots of sugar out on 5-day patrols in Africa. Tea and sugar take no carrying and the water you need to drink anyway. It was certainly better than rancid corned beef which was the only alternative. An added bonus is that it is easy to make.

Nowadays there are energy drinks available which will give you a quick burst of energy and this can be useful in the short term. The drinks also often give you a shot of caffeine which can be a help when you are tired. I know there are arguments against the use of caffeine as a stimulant but most people will agree it is better than dozing on patrol or guard and getting the back of your head blown off.

The problem with energy drinks is that they come in cans and cans are heavy to carry about in addition to water until you want to use one. A good alternative is to buy the energy drink in dried form – or better – make one up yourself from glucose (a 'simple sugar' which is easy for your body to use), caffeine and flavouring. Keep it dry in a bag and add water when you need it.

Food dropped wherever you need it – tasty treats being rolled out of the back of a Hercules C-130. (Photo courtesy Sergeant Roy Mobsby)

# LIVING OFF THE LAND

We all know the modern soldier is provided with all manner of tasty treats in his rations – don't bullshit me – but if you are Special Forces then you may be away from re-supply for an extended period and find it necessary to find your own food. This is not 'survival' training remember, so not exactly living off nature's bounty, but I will tell you here how to stay out on patrol longer when lack of food would otherwise force you back to base.

The first thing you need to realize is that most of what you see on TV and read in the survival books is bullshit from a soldier's point of view. True woodsman-type survival skills take all your day, every day, to scrape a living together in the form of water and calories. This is no good to you while you are trying to fight a war. What you need is a quick supply of water – which we have seen you can get from the locals – and a handy supply of food rich in calories.

The sort of vegetable food you can find in the wild, even if you know which to eat and how to prepare it, is generally so low in calories that it is hardly worth eating. As a general rule, green plants don't have much in the way of calories. Calories come mainly from the roots which supply the plant with the energy to grow next year – like potatoes and yams – or from the seeds such as corn and many types of nuts. As these don't commonly grow wild you are not going to get them except from other humans.

A British Pathfinder patrol stops in a village surrounded by turkeys, Afghanistan 2010.
Don't steal the locals' food unless you are on the run and desperate. They have little enough as it is and it will never win you any favours. (Photo courtesy Tom Blakey)

## Food from the locals

Very often the locals will be pretty poor. Civilians generally are poor where a war is happening as they maybe can't work for landmines and the insurgents probably rob them. So the chances are they will sell you food if you produce a few dollars. It is a fortune to them and nothing to you. But there are down-sides to buying your food from the locals.

If you are on a covert mission then you can be sure the locals will tell the enemy you are there. If you have to talk to the locals keep most of your men hidden so they don't know your numbers and make sure you set up a fire position outside the village and keep your eyes open while the buyers are in there. I have been hit while doing this as it is so easy for the locals to keep you talking while they send for the enemy to come and ambush you as you leave. This way, remember, they get your dollars and a reward from the enemy for ratting on you.

Grain and fruit is pretty safe from local suppliers but make sure they don't have time to doctor it before you make your deal. You don't want any nasty additives such as pesticide.

Meat is a little more interesting as, dead or alive, it is more likely to have worms, liver flukes and other parasites than when supplied from your family butcher. Thorough cooking will render these safe and they add protein.

## Catching your own food

The only wild foods worth collecting are meat and fish as these are full of calories and fairly quick to catch and prepare. If you are fortunate enough to be patrolling near a lake or river then setting baited lines will very likely produce a pleasant addition to your diet if left over night. Meat depends very much on where you are: goats and deer are a favourite as a compact source of meat and may be shot or trapped as required. If there are rabbits or other furry creatures around then snares are the easiest way to catch them.

Kudu shot by the lad in the shorts from the turret hatch of a moving armoured car. Author third from left in greasy coveralls and all of us still a month from a bath. (Author's collection)

## Tasty treats

You might want to carry a light-weight treat such as curry powder or other strong flavour to kill the taste of food which is past its sell-by date. Curry powder doesn't come from hot countries for nothing and slightly rotting meat won't kill you – whatever the Health Police say. Always cook meat thoroughly to kill the parasites and then add curry powder to kill the taste. If you are ever going to live off the land carry salt – there is nothing worse than fresh meat without salt – well there is, but you know what I mean.

## Eating on the run

If you ever have to steal meat and run – say you were escaping from somewhere – kill the animal by cutting the carotid artery to the side of the neck and hack off one of the back legs. This is more meat than you will need. Stuff the leg down the front of your shirt to carry it. If you need to keep the meat for a while in a warm country cut it into strips like boot laces and hang it on bushes to dry. Dried raw meat is called 'Jerky' in some parts of the US and 'Biltong' in Africa. This way it will last for months rather like Parma Ham which the Italians prepare. It's quite tasty and very nutritious – plus it gives you something to chew on so you feel you've had something.

# COOKING IN THE STICKS

If you are eating ration packs then there is no need for a fire except to keep warm and raise morale if the tactical situation allows: you do make a marvellous target for a sniper huddled round a cooker or camp fire. Don't cook at night – I would say ever but I know you will. Walk off a little way from a mate sitting over or by a fire/cooker and see what a good target he makes and then think about it. Sometimes you can dig a little hole to hide a cooker but it will still light up your face as you look over it if you aren't very careful.

Given you are in a situation where you are cooking food you have stolen, bought or caught then you are likely on a Special Forces long duration mission and so you may also be building shelters. Where you put your fire relative to your shelter is quite important tactically and for getting the most warmth out of it so I will cover the cooking aspect of fire here and then its siting in the section on shelters.

## Preparing fish for cooking

Fish can be eaten raw — few Westerners knew this before sushi became popular in recent years. Your body can get lots of nourishment out of fish without it being cooked but fish flesh can be full of worms and cooking kills these. A few worms in your belly for the length of a patrol will do no harm and they are easily removed with a de-worming pill when you get back to base. Just like you give the dog. Worms, contrary to urban myth, don't make you thin by stealing a substantial part of your food. What they do is irritate your gut which can make you nauseous and so you eat less.

Cut off the head and tail of your fish. Save these for bait if required. Slice along the belly and drop out the guts. Cut further into the spine all along the length so the body can be opened out flat. Fish cooks quickly in any event but this flattening allows for even quicker cooking. You can cook over a fire by hanging the whole body lengthways along a stick. As an alternative you can fold the body back together and place on a hot rock or in ashes wrapped in a large leaf. The skin then holds the meat together a little. To pad out the meal chop the fish into pieces and boil with corn, veggies and curry in a pot. If you don't have a fire then prepare the fish in the same way but dry it in the sun to keep for weeks.

Gutting a fish is made easier with a decent knife. (i-Stock)

## Preparing meat

Meat can be eaten raw but it is more difficult to digest than fish and you will get more nourishment out of it if you cook it first. The heart, kidneys and liver of, say, an antelope are particularly good for you but check the liver does not have purple lines or other markings on it as that is a sign of liver fluke which is not good for you. Discard a liver with these signs and the remainder of the carcass is still fine to eat.

**LEFT**
Ace machine gunner and very passable butcher Musa prepares a gazelle I shot while on patrol. (Author's Collection)

**ABOVE**
The first step to butchering a deer or large buck is to remove the skin from the entire body. (i-Stock)

The most convenient, if not quickest, way to prepare a deer-type carcass is to make a cross-bar of a pole at a suitable height over a hole. Then you hang the dead animal by its back legs from the bar. First remove the head. If the animal is freshly killed the meat will last longer and eat better if you allow the blood to drain from the body before it has coagulated (i.e. gone thick). You may want to bake the head and extract the cooked brains as a treat. Or you may not.

Next remove the skin by making cuts around the 'wrists' of each limb. Continue each cut up the inside of the limb then join them with a cut along the belly which runs to the neck. Be very careful not to cut too deep as this will burst the membrane which holds the guts and result in their contents of shit and bile squirting everywhere with a foul smell which will taint the meat.

Tease a corner of the skin away from the flesh at the wrist and then peel it from the body all over. This is a lot easier to do than it sounds. Though the idea of keeping the skin as a souvenir may appeal to some, it needs treatment to stop it rotting or going stiff as a board when dry.

Now, very carefully, cut from the anus to the throat going just deep enough to cause the body cavity to open up but not deep enough to open the intestines as mentioned above. At this point the guts will start to come out of the body and should be allowed to drop into the hole provided for this purpose. You will have to release them from the body at one or two places including the anus and the diaphragm. When this has been accomplished you will have a nice, blood-free carcass ready to be cut into pieces for cooking.

Take off the legs first by cutting a circle around the top of each then carefully finding the ligaments which hold the joint in place and cutting them. There are a whole range of ways to dissect the remaining body depending on what you want to do with it. The meat running along either side of the spine is the fillet and a favourite cut.

A sizeable animal not only produces rather a lot of meat but the meat is in large, thick pieces and so takes a long time to cook. If there is more meat than you will eat in a day or two then, if the tactical situation and weather allows, you can preserve it by cutting into thin strips and drying on hot rocks or bushes in the sun.

There are many ways of cooking meat with a fire but small pieces on sticks, or larger pieces buried with hot rocks are convenient if you are short of equipment. Probably the least wasteful method though is to cut the meat into inch-thick cubes and boil in a pot with water and vegetables.

## TOP TIP!

### Dealing with your waste

Bury your waste deeply as the local wildlife will want to check it out for leftovers and depending where you are this may be a problem. Also, rotting food will attract even more flies than your usual allocation.

## Cooking support

There are two easy ways to make a support for your cooking pot if the local vegetation allows: a tripod over the fire with a thinner stem hanging down and using a side-shoot from that as an adjustable hook is the best. Failing that one pole hanging at an angle over the fire wedged at the ground end with stones works well enough.

## Brewing tea and coffee

A regular supply of tea or coffee, depending on where you come from, does wonders for morale in uncomfortable postings or positions. As an officer or NCO it is a good idea to ensure the men get a brew at regular intervals when in a static position. On patrols even, a brew at a break helps the day pass a lot easier – we are not trying to make life hard work after all.

# CLOTHING

What do you wear clothes for? OK, I know, you don't want to frighten the ladies. There are several other reasons you might want to wear clothes on operations – these are to keep warm, keep the sun off, stay dry and not be seen.

## Cold

Cold kills as surely as a knife. I have always been a warm-weather soldier: I just don't like the cold and have always felt that if someone wants some place that is really cold then they can have it. Joking aside, unfortunately, I have had to spend time in places a lot more cold than is pleasant. I know soldiers who have had to cut off their toes to stop frostbite spreading. Keep yourself warm or you will never tap-dance again.

If you find yourself in a cold climate then you have an interest in keeping yourself both warm, dry and out of the wind. You should be issued with warm clothing but if, for some reason you are out without the right kit then you can improve your situation by packing the space between your trousers and legs and jacket and body with paper or straw. This will insulate you to some extent. As a note I will say here that you want tight trousers and jackets to look smart on the

**REMEMBER:**

If it gets very cold, say below -20°, then watch out for touching metal objects. Because metal conducts heat so well it conducts the heat away from your skin and causes the moisture to freeze sticking you to the object be it rifle, vehicle or whatever. I have seen it happen in a German winter and it can be very funny to watch but somewhat inconvenient to experience. Don't let it happen to you.

parade ground and very baggy ones for soldiering in. Baggy clothing both breaks up your visible outline and gives you room to carry gear inside or wear the clothing over bandages and dressings. I often slip empty mags into my jacket front as I take them off the weapon when time is an issue.

## Wind

Whatever you do, wear windproof clothing in cold weather. I have worked in Europe where it has been 30° below freezing and often it has felt OK when there has been no wind. But with a wind just a few degrees of frost can feel like Siberia.

## Rain

If you get wet and the weather is bad you will be cold and miserable, especially if there is a breeze to carry your heat away. The thing about getting wet on patrol is that you often cannot get dry for days. Your clothing rubs up sores and the skin peels off your feet like you have some terrible disease. This makes life a lot less pleasant and can be avoided to some extent.

If possible, when you have got wet, and you are in a safe area or well hidden, take off all your wet clothes and get into your sleeping bag with them. The bag will soak up the moisture from your body and you will soon be warm and dry. Eventually, your body heat will dry out the clothing in the bag with you. Be warned – putting on cold, wet clothing in a cold climate is not for the faint hearted.

See how this gunner blends into the bush background wearing just his Rhodesian uniform. (Author's Collection)

## The poncho

A long-time favourite of mine has been the 'poncho'. The name comes from the square of material with a hole in the middle worn with the head through the hole like an all-round cape by the Spanish, the Mexicans and Clint Eastwood.

The sort of poncho I am speaking of is made of a waterproof, camouflaged fabric. It consists of a rectangle of material about 2m square with a hood fixed to the hole in the centre. Worn as intended it keeps the rain off better than almost anything without causing you to get sweaty because the breeze can still get around you. You can wear a pack and carry a rifle ready for use under the poncho and keep these dry too. Then you can use it for a tent, a sunshade, a windbreak or a groundsheet.

## Sun

A strong sun is a killer too. Always wear a hat which casts shade on your neck whenever you are out in strong sun. If you get the sun on your neck for a while you will get sunstroke. You will only do it once as it feels awful. You will at best feel dizzy and as sick as a dog. Get it bad and you are out of action for days. This can be fatal if you are out on patrol without support.

Some people prefer a broad brimmed hat and others prefer a 'kepi'. For the non-French amongst my readers, a kepi is a hat of the style issued to the French Foreign Legion. Its significant feature is the neck flap which hangs down behind like the tail on a 'Davey Crocket' hat. Think of a baseball cap with a flap behind to keep the sun off your neck. The broad brimmed hat looks less smart but does keep the sun off your ears.

Men who have worked in Africa for a long time tend to cover all their body from the sun. I wear a kind of camouflage overall – coverall in the US – in hot countries as it keeps the sun off and is loose round the wedding tackle to avoid sweat rash.

> ## REMEMBER:
> Guys with a lighter skinned European ancestry will burn quicker but no matter the colour of your skin *everyone* is susceptible to sunburn and heat stroke.

## Heat

When you move to a hot climate from a cold the heat gets you down for a while and you feel tired all the time. There is nothing you can do but put up with it for the couple of months it takes for your body to adjust. Drink plenty and take the salt pills while you are waiting. After you have become acclimatized, and it takes about 4 months to get fully acclimatized, the hottest weather will not bother you.

Notice how Arabs and Africans wear heavy clothing in the hottest weather. This is because they are accustomed to the heat. Temperatures like an oven can feel cool to them. You will get the same providing you stay in the heat and don't spend all your rest time in air conditioning. The main effects from heat are feeling as though you don't want to do anything and a true reduction in your physical performance – which is quite different. Make sure you stay alert when it matters – bullets accept no excuses so you need to use that old-fashioned concept of will power.

When I first went to Africa as a youngster I worked out in the sun with my shirt off for a couple of hours. It only felt like a warm day and I have quite dark skin. The next day I was covered in huge blisters everywhere above the waist. I had to complete forced marches in this condition and my shoulders bled from the straps of my pack. I have been very careful of the sun ever since then and I want you to be the same.

## Camouflage

The idea of camouflage pattern uniforms is that at a distance you should be invisible. This just doesn't happen in practice but wearing camo does make you less obvious. To get really invisible you have to add some of the local plant life to your outfit.

Have you noticed that all the different countries involved in Afghanistan all wear different camouflage pattern for the same job? This just confirms my belief that for day-to-day use – where you are not, say, lying up in an observation position or as a sniper – so long as the camo is dull it makes little odds. So wear your national camo and make sure there are no shiny bits showing to catch the sun and draw attention to you.

# DIY SHELTERS

Humans are made to operate pretty well in what we consider to be a mild climate without too much sun or rain. As a soldier on patrol the problems you face are too much sun, heat, cold, wind and rain. The places you are going to find the inclement weather are mountains, deserts and forests. Mountains are cold because they are high up. They are always windy for the same reason. Deserts are hot in the day, mostly, and cold at night. Forests tend to be wet – or they wouldn't grow and they are often cold too. If you are so inclined, you can buy

Home-made shelter. It is unlikely that you will have the time to make a shelter like this while out on a mission, or that it would stay undiscovered for long if you did. (USMC)

a survival book for the type of area you are posted to and, like the boy scout, be prepared. In case you can't be bothered here are a few ideas for making yourself comfortable.

## When might you need to make a shelter?

You might think that the only time a soldier needs to worry about shelter is when something has gone seriously wrong – like you are on the run in enemy territory. Actually, if that is your situation you are very unlikely to have time to make a shelter unless you are having to lie up while an injury heals.

In the normal course of events while on operations you will be issued with equipment suitable for the climate. There will be buildings or portacabins of a suitable type for when you are in your base camp and your kit for patrolling will include clothing and sleeping bags of a suitable grade for the climate. Of course they will... So when do you need to make a shelter?

I have said before, and I will say again, being a good soldier is not about roughing it. The idea is to make yourself at home wherever you find yourself. So, if you are dug in to a defensive position on a hill, in a forest or maybe even just waiting for a few weeks in the back of beyond for the generals to make up their minds what to do with you then you might as well be comfortable. Being comfortable means being out of the sun, cold, rain, snow or wind.

And, of course, the tents will have gone missing. Depending on the situation, your shelter from the weather may also have to double as a shelter from artillery and gunfire but I have covered that in another section so just take this into consideration if it applies.

## Degrees of shelter

- **Clothes:** The basic shelter is the clothing you stand up in — given you are wearing the right gear it should keep off the sun, rain or whatever. But it needs to be light enough to walk and work in so maybe when you stop for a break it gets chilly.
- **Sleeping bags:** The next level is the sleeping bag. These should have a ground sheet built in so you can throw them on the ground, climb in and pass out if the opportunity arises, and as long as it is not Arctic conditions.
- **Poncho wrap:** If you have a situation where strong sun is a problem then clearly you need some type of sunshade —a ground sheet strung between a couple of trees or poles set up for the purpose is just the thing.

   But when there is rain, snow or cold wind you need to start thinking. The first thing to realize is that wind not only carries away your body heat and makes you feel colder but it also drives rain and snow into your clothing. This is a bad thing so you must first of all protect yourself from the wind.

   The quickest and easiest way to do this, at a night stop in bad weather or to avoid putting up a tent, is to lay out your old friend the poncho with your sleeping bag along one side. Get in the bag and then roll up the poncho around you. Given you don't sleep where ground water runs into the envelope you will stay snug and dry in the worst weather. This is all very fine for a quick sleep but if you are stuck in one place for a while you don't want to spend all your time in a sleeping bag. You want to cook, eat, smoke and so on. This means a more elaborate shelter.

- **Poncho tent:** The next level is to put up a ridge pole between two trees and sling your poncho over to make a lean-to or other type of roof with the wind behind you. Your buddy's poncho can make the groundsheet if necessary. You can then both sit in comfort drinking hot tea and smoking for hours in the heaviest rain. Listen to the radio and oil your rifles if you have nothing better to do. At the opposite extreme, ponchos make good sunshades too. Sitting in a hole with the sun beating down can be a tedious experience and a poncho, a few

GPMG and Bren Gun positions, 200 miles north of the Arctic Circle. Note the contructed shelter. (Photo courtesy Colour Sergeant Trevor 'Sadie' Sadler, 1st Battalion 'The Vikings' Royal Anglian Regiment)

sticks and a bit of string can make all the difference to your quality of life. For all this, the poncho weighs a few ounces. If you cannot buy one locally then make one from plastic sheet. Put eye-holes around the edges to make attaching guy ropes easier.

## Local materials

A poncho and/or some plastic sheeting weighs nothing and should always be in your kit when there is a chance of staying out anywhere. Failing this, or when you need something more substantial for a larger refuge, take a look around for suitable materials. If you are in an area which is or was built-up then there may be corrugated iron. This is easy to work with to make a wind break or roof and when placed on strong timbers can support a lot of protective earth.

Failing heaven-sent corrugated iron you are down to making a wooden frame and covering it with plastic or even thatching. Thatching with grass or other shrubbery is rarely rainproof but your first thought should be to stop the wind; then the rain from above. Even slowing the rain to a steady drip is much more comfortable than driving rain as it is the wind which will do you down.

## Heating

So you realize now that all accommodation should have its back to the wind, so to speak. Generally this means a sort of lean-to with the open side to the lee of the wind. It is possible, if the tactical situation allows for a fire, to heat this quite efficiently. Site your fire in front of the opening on the lee side of your shelter and perhaps 4 or 5ft away. Drive in stakes to make an arc on the far side of the fire and then lace these with thinner shoots or whatever offers to make a curved wall behind the fire from where you are sitting. Surprisingly, this sort of screen reflects the heat from the fire quite nicely into the opening of your shelter. Being close by, it also means you can sit and watch your food cooking while you are out of the rain.

ARVN's (Army of the Republic of South Vietnam) shown digging bunkers and piling sandbags around their fire support base Delta One about 9 miles inside Laos in 1971. Nice position, nice trenches, nice shelters, good effort. (Corbis)

## The dangers of smoking in the field

Soldiers smoke to deal with the tension and boredom of the endless waiting. There is an argument that says if you may well be shot soon you might as well risk cancer from a smoke.

But, remember, a good tracker can follow a smoker by the smell – even when he is not smoking. Certainly he can lead a team in to a camp where there are smokers from down-wind. This may or may not be a problem depending where you are and the soldiering skills of the enemy.

What is always a problem is the light given off by the cigarette whenever it is lit and glowing, but especially when the smoker takes a 'drag' or actually strikes the match. When I was a 16-year-old junior soldier we were taken out onto some cold, dark, windy moor in the north of England to watch someone light a cigarette at night away from local sources of light.

It was quite amazing how much light came from the lighting of the cigarette, particularly, and how the glow from the cigarette lit up the smoker's face as he enjoyed his smoke. By adjusting our distance from the target we worked out we could see the smoker light up from some hundreds of yards away! I have always remembered this and send my thanks to the instructor if he reads this.

In the British Army there used to be a tradition that the 'third light' for a cigarette from one match was unlucky. This dates from the Anglo-Boer War (1899–1902) when crack Boer snipers would see his target with the first light, set his sights with the second and make his kill with the third.

# CHAPTER 4

## Staying Healthy

Your health is another of those things you don't think about until it goes wrong. Especially when you are young and fit. It's pretty obvious that if you have a leg injury and can't walk then you can't soldier. If you have an arm or hand injury you can't hold a rifle either. What is not so obvious is that an illness, such as where you have to run to the latrine every ten minutes, also takes away from the alert frame of mind which makes you a useful soldier. Your priority then, to stay alive and be at least some use to your mates, is to stay as fit and healthy as possible, given the conditions, so you can keep your mind on your job. Here we will take a look at what you can do to achieve this.

# MILITARY FIRST AID/TRIAGE

You can stub your toe or cut yourself shaving just as well when you are soldiering as a civilian can back home. The difference in your job is that on top of all the usual bugs, bumps and bruises you might collect as a civvy, you are also going to be faced with a range of interesting injuries caused by blast or pieces of metal entering the body.

Most wounded soldiers die in the 15 minutes they are waiting for a chopper with medics aboard. What you are trying to do with your first aid, or triage as our US friends call it, is keep your mate alive until a helicopter comes to take him to a real medic. Most casualties don't die of being blown to bits or shot through the head – they die of choking on their own vomit while unconscious or of blood loss because they are not patched up properly. Both of these are easy to deal with. A child could do it so make sure you can.

Casualties in Afghanistan coming in off the chopper. (Dave Husbands © UK Crown Copyright, 2009, MOD)

Before helicopters, those wounded in combat used to die in droves. Depending on the war and time period, you stood a better than 80% chance, if you were wounded seriously, of dying from something simple like blood loss, shock or infection. Starting in the Korean War the US Army started to bring the wounded out with helicopters and this cut fatalities dramatically. Nowadays, if you are wounded seriously, you stand an 80% chance or better of survival. Keep these figures rising by learning how to patch your mates up. You only have to keep them alive for a few minutes to make all the difference.

## REMEMBER:

Your knowing how to do the necessary will save your mates and your mates' knowledge will save you. When there is blood and snot everywhere is too late to pay attention to the first aid lectures. And when your mate is squirting blood all over you it is a poor time to freeze up because you don't know what to do.

When you see someone hurt it is often obvious what has happened but when it is not you check things that will kill him quickest first:

- Check the heart is working by placing your ear to the chest and if it is not give it a bump to start it. Lay the casualty on his back and strike down firmly over the heart. Place one hand over the other and compress the chest sharply with the heel of the lower hand. Listen again to check the heart has started. If it has move on to breathing, if not strike again.

- Check the breathing is ok and if not clear the airway of puke or teeth and then give artificial respiration. Your *real* mates will do this. Inflate the chest by holding the nose and blowing into the mouth. Allow the chest to empty and repeat until the casualty takes up the stroke himself. A bullet wound to the chest will often collapse the lung as the lungs are naturally filled when you breath in by the diaphragm pulling down and expanding them by suction. When there is a hole in the chest cavity, such as may be made by a bullet or shrapnel, the lung fails to open as air comes in through the hole. What you have to do is seal the holes front and back with plastic and put the casualty on their side with the wounded lung down so it doesn't flood the good one. People can breath OK with one lung so long as they don't try any gymnastics.

■ Stop any heavy bleeding by applying pressure to a pad placed over the site of the wound. Providing it is a surface wound this will stop the bleeding nicely – even where a bullet has gone all the way through a leg or other limb – but where the wound is in the stomach or chest the body will bleed internally. The casualty is then dependant on the medics fixing him up in the 30 minutes before he bleeds to death.

Pressure on a wound is generally more effective than a tourniquet in most cases as a tourniquet can cause the loss of a limb below the constriction if it is not relieved regularly. A tourniquet is a band or rope or similar pulled tight around a limb. Tight enough to stop the loss of blood from a stump or leaking artery – and these are really the only time a tourniquet should be used. Tighten the band until the bleeding stops. This should not be too painful for the casualty but they can have morphine. Every 15 minutes release the pressure to allow blood to flow to the limb and avoid it dying. This release of blood is quite painful for the casualty so expect complaints.

Give an intravenous drip to all casualties as the saline solution will both counter-act blood loss and help with the physical effects of shock. This is essentially a bag full of salty water

American soldiers teach Afghan police basic first aid. (USMC)

with some minerals in it. Usually there is a long tube with a needle attached to deliver the fluid into the arm. Stick the needle into a vein in such a way that the end of the needle stays in the middle of the vein and allows blood to flow freely. Hold the bag high and run a little solution through the tube to get bubbles out of the system – bubbles in the line can cause a fatal clot in the body. Connect the tube to the needle and adjust the flow to a steady drip through the measuring device on the tube. Getting the needle into the vein just so takes a little practice.

We played for hours practicing on each other during my time in the Bush Wars in Africa. When the needle goes in it will often go through the far side of the vein wall and then you don't get a proper flow – you get a big lump on your arm. It's all down to the angle of attack – and practice. As most medical students will know, drips are also an instant fail-safe cure for hangovers but they should only be used by soldiers after training and in an emergency situation.

## REMEMBER:

Clinical shock is not just a nasty surprise. When a body is injured it engages the natural survival reaction to withdraw blood from the extremities and concentrate on keeping the major organs going. This is not always the best thing for the body as it can cause the temperature to drop and various other problems. The point is that a casualty should be kept warm as a remedy for the effects of shock because medical shock on top of the injury can kill.

## Recovery position

When you have patched the casualty up to the best of your ability you have to decide if he is in a fit condition to turn over into the 'recovery position'. This is three-quarters face down with the upper knee drawn up towards the chest and the upper arm supporting the chest. The idea is to have the face pointing down so as to allow the airway to clear automatically should the casualty be sick. Many unconscious casualties choke to death so if he can't do the recovery position, owing to his injuries, keep an eye/ear on his breathing.

By these simple procedures you can keep nearly anyone alive for a while – long enough for someone who knows what they are doing to get there.

## Pain killers

Soldiers have carried morphine in combat for years. It usually comes with a sort of applicator so you can stick it in easy — your own leg or someone else's. Morphine takes the edge off and stops screaming but it should never be given to someone with a head injury.

## Antibiotics

In warm countries especially, cuts and other wounds get infected easily. What would be a simple matter to take to the medic in camp can be fatal in the field. If you don't have a specialized medic with your team carry a course of Broad Spectrum Antibiotic (BSA) pills, carry a syringe and the liquid form too if you can get it.

Antibiotics work by killing the 'bugs' that make the infection. I have found that the best way to take antibiotics is to inject a good dose into any muscle, so you get a decent level in the body straight away, and then follow up with a course of pills to maintain the level required to kill the infection and bugs. Find out from wherever you get the BSA how long the course needs to be. It goes without saying that you should never dose yourself, unless emergency medical support is not available.

That is about as much as you need to know to be a bit of help to your mates when the shit is flying. Do it right and you will make a friend for life and feel warm all over. This is worth learning properly and practicing so make sure you get yourself, and all your mates, on a good first aid course before you go into combat.

# PERSONAL HYGIENE

We have already seen that you can function best as a soldier if you are healthy and relatively happy. By and large you will feel better if you can keep yourself clean and healthy. Certainly you will be more popular. Sometimes you can't and then you just have to put up with it.

Generally, when it is cold you might not wash because you don't want to take your kit off. When it is hot you don't wash or shave because there is no water. Being dirty in the cold might not do much harm, other than to your social life, but failing to look after yourself where it is hot can have serious results. Hot climates are where all the bugs and other nasties live. By the nature of things, therefore, most of what follows is more important in a hot climate.

## Shaving

If you are working undercover or spending time talking to the locals in an Islamic area you may have to grow a beard. This might be to avoid attracting attention and you may not be taken seriously without it. If you have not grown one before you will find they itch for a while when they are getting started. And they itch a whole lot more when they are full of sweat, dirt and bugs after a few weeks out in the sticks. That is why Western soldiers shave.

In the British Army shaving used to be mandatory for everyone no matter where you were. I remember in training living in muddy slit-trenches on snowy mountains in Wales but we still had to be properly shaved every morning. We were taught how to save a drop of water in the corner of a mess tin then to use the corner of a towel as a flannel to wash and then take a shave with the remains. It probably was good for exercising self-discipline and maybe even helped morale but I think it is a total waste of water and time on active service.

## Hair care

Hair is best kept cropped short for coolness and to avoid bugs and rashes. If you leave long hair for a month or so it becomes self-cleaning. I have experienced this myself many times as a young man – when I used to have hair. For the first few weeks your hair gets greasy then it seems to dry and stabilize. Animals don't get washed in the wild and we didn't evolve to wash either. This must be natural.

## Brushing your teeth

A tooth brush doesn't weigh much and it takes little water to clean your teeth. Most of us feel better with clean teeth too. I confess I never used to clean my teeth on patrol in certain areas as it was a consideration that trackers could pick up the smell. I suggest that where tracking is not a problem you clean your teeth. At the least this will avoid gum disease which could be unpleasant as well as costing you your teeth.

## Keeping your camp clean

Where you don't have proper latrines try to bury or cover your crap even if you only stop for the night. If you stay for longer it is absolutely vital. Flies are attracted to crap and walk on it. They then use the same feet to walk on your food. This is not only gross but the flies pick up diseases such as cholera from their habits and you can easily catch it. Local water resources can also easily be contaminated by sewage.

It is probably not as well known that hyena steal crap. They creep up at night and eat it if they can. I'm no expert on hyena but one night we were free-wheeling in an armoured truck for some miles down a long hill. We knew the enemy would see and hear the truck intermittently from their observation positions so the idea was that we would jump out without the truck stopping and it would then go on to another base. We could then creep up on the enemy. All of a sudden this 20-ton truck gave a lurch. We pulled in and found we had hit a huge hyena and killed it. Fortunately, as it was like a bear it was so huge! The head was enormous. People don't realize that hyenas hunt game themselves and will drive lion off a kill. Their jaws are said to be the strongest of all mammals and can crush the thighbone of an ox.

The reason I mention hyena in depth is because a guy I knew was sleeping with his arm out of his bag one night and a hyena was attracted to camp for whatever reason. It crept up and bit off his hand and ran away chewing. The moral of this little titbit is don't leave your waste lying around. In Africa it would attract hyenas, in other parts of the world it could easily be something which is equally as dangerous.

## Sweat rash

Some people are more prone to sweat rash round the crutch than others. I don't know why. If you get it carry a suitable powder or cream. Left untreated and unwashed it can get nasty — your bollocks will rot and drop off.

When I used to go on long operations — a month in the sticks at a time — I used to wear a camouflaged overall or jump suit. Though eventually they used to go black and greasy, they were loose around the waist and crutch and worn with no underwear this cut down on the exposure to sweaty cloth which nurtures fungus.

# INTERESTING DISEASES

You can catch a cold somewhere exotic just as well as in London or New York but when you travel to exotic parts they have exotic diseases and your body's defences are just not as ready for them as the locals'.

A good example of a clean camp, with washing drying in the midday sun and a cover to keep the dust out of the mortar tube. (Dan Bardsley © UK Crown Copyright, 2009, MOD)

Of course, some diseases are more common in some places than others so make it your business, if your attached medic hasn't already, to find out what the locals' favourite diseases are. Your best bet is to just keep an eye on your general wellbeing and don't be shy about visiting the medic if you have any amusing or unusual symptoms. Worldwide, the most popular illnesses, and by far the biggest killers, are malaria, HIV/AIDS and tuberculosis.

### Malaria

You get malaria from mosquito bites and cannot avoid these on operations so take the pills provided by your medic. Catch malaria and you will regret it as it starts like flu and gets worse with jaundice, diarrhoea and vomiting. There are several types and some will kill you despite medical intervention.

### AIDS

Soldiers get AIDS from whores. When I was a boy you couldn't catch anything a jab in the arse wouldn't get rid of but now you can catch your death. Keep it in your pants if you can, otherwise if you insist on being stupid wear a condom. And **never** kiss a whore. This is your dad talking.

## Tuberculosis (TB)

Get the jab for TB as it starts as a cough, then you cough up blood, then you die.

## Amoebic dysentery (Amoebiasis)

Amoebic colitis is caused by the parasite Entamoeba histolytica which lives in the water. Use the pills supplied to purify water and you will avoid the squirts and abscesses in your guts which hurt a lot. Dying from amoebic dysentery is a messy business.

## Chagas disease

Chagas disease is very popular in South America. It is the result of infection by the parasite Trypanosoma cruzi. This little creature lives in triatomine insects or assassin beetles – an insect up to an inch long which lives in crevices in houses. These beetles creep out at night and suck your blood leaving the parasite in exchange; but you can also catch the parasite from eating food they have crapped on. Chagas disease starts with swelling, often around the eyes, then you get stomach problems, your spine and brain swell and eventually you die.

## Cholera

Cholera is caused by bacteria ingested through food or water contaminated by sewage and it is extremely easy to catch if it is about. It causes severe diarrhoea and vomiting, which can lead to dehydration and death.

## Dengue haemorrhagic fever

Dengue virus is transmitted to humans by mosquitoes. There are four serotypes of dengue virus and they cause a wide range of symptoms from mild febrile illness to dengue haemorrhagic fever. This is a severe illness with fever and bleeding from all your holes. Left untreated it leads to dengue shock syndrome which, unsurprisingly, comprises clinical shock, collapse and death.

## Leishmaniasis

This is the name given to a group of diseases caused by protozoan parasites of the genus

Leishmania. The parasites are transmitted by bites from infected female sand flies common in many tropical and subtropical countries. These are not actually flies and they don't live in sand. Though symptoms vary, the nose, throat and mouth get sore and swell first. Then your spleen, bone marrow and lymph nodes swell and stop working with the obvious result: you die.

### Trachoma

Trachoma is a bacterial infection in the eye caused by Chylamydia trachomatis and is common in the poor communities of low-income countries, mainly in Africa, Asia and the Middle East. Through repeated infections, the eyelashes turn in and brush against the cornea. The contact between the lashes and the surface of the eye gets sore and results in blindness. The good news is that you are going to seek treatment for this when it hurts and before it makes you blind.

I think that is enough to make you keep an eye on your health. OK?

# BITING CREATURES

The animals to be afraid of are not the big cats, the alligators you meet on jungle exercises and so on. Insects are by far the most dangerous creatures when you are a soldier.

## Flies

In the West everyone knows that flies carry disease yet no one seems to catch anything. In warm countries flies do carry lots of disease and people do die – indeed, people die like flies. You can get fly traps for camps and fly repellent for personal use but flies will still find you. Try to keep food and refuse out of their way and, if the flies are driving you crazy, wear a hat with corks hanging like Crocodile Dundee – seriously, they do work!

There are some kinds of flies which bite and carry disease. There are others which come and lay their eggs on or in you and the maggot eats it way through your flesh. Some eat their way right through you to emerge from your skin when they want to turn into flies themselves. This is unfortunate if they choose, as some prefer, to exit through your eyeball. If you think you have something eating you see the medic and it can be removed immediately.

## Mosquitoes

These little creatures are clever and deadly! They kill more people – millions – every year than all the wars and accidents combined by the spread of malaria. Wherever you go in a warm climate you will come across mosquitoes and you will get bitten.

The way a mosquito operates is as follows: the pregnant female – the only sort that drinks your blood – operates at night, sensing your scent or heat from some great distance away and homing in on you. She picks the warmest bit of bare skin she can get at because there is more blood and then she sets to work. Using her very fine proboscis she injects anaesthetic and settles down to a pleasant meal of your fresh blood through her 'straw'.

Unfortunately, her visits have two unintended side effects. The first is that the chemical she injects to stop your blood clotting also sets of an allergic reaction in some people. The bites come up in puss-filled boils and the 'bitee' feels like death. The antidote is an anti-histamine but don't take more than the stated dose or you will feel even worse.

The other potential side effect is malaria, a disease which kills millions of people every year. You don't want to catch it as, at best you will have attacks of fever and cramps at irregular times for life, and at worst you will die messily.

There are two ways of avoiding malaria, you either stop the mosquito biting you or you take a drug to stop the disease taking hold. While you can avoid the majority of the unpleasantness from mosquitoes through the use of netting and sprays you cannot stop them all. You will get

**TOP TIP!**

### How to avoid interesting diseases...
- Before you go on operations get the correct jabs
- If the area has malaria start taking the pills in advance if the medic tells you to
- If operational security allows, use bug spray and sleeping nets
- In camp spray the room or tent with bug spray regularly
- Keep your accommodation clean with regular brushing
- Wash dirty clothes and bedding when you can
- Dispose of food and human waste where it will not contaminate your water supply or attract scavenging animals or flying insects

bitten. So the only sensible thing to do is take the pills. Your medic will provide some little pills you take every day. Like showing a smoker someone dying of lung cancer, perhaps showing a soldier someone dying of malaria might convince them to take the pills. I can't do that but maybe your boss can?

## Tsetse flies

These little pests carry sleeping sickness. A victim feels like they want to do nothing but sleep. Tsetse flies hang about during the day on dead leaves or other vegetation. As you walk past they fly up and bite you like a horsefly. It feels like a jab with a needle and comes up in a little lump. As you walk carrying a pack they tend to bite on the back of the shoulder where you can't brush them off. Fortunately, you only get tsetse flies in Africa. A sure sign of a Tsetse fly area is the lack of livestock as the flies kill all the cattle. If they are likely to be a problem your medic will supply the appropriate drugs, but always voice a concern if you suspect you have been bitten.

## Lice

These are harmless little critters that live on the bodies of people who don't wash too often. If you spend time with the locals in some areas you will pick up a few lice. Your friendly medic will give you a shampoo or similar for instant removal. The real problem with lice is social — you may clear a space at the bar.

## Ticks

Ticks are little insects which bury their head in your flesh and suck your blood. The blood tends to fill up the body of the tick which appears like a sack the size, perhaps, of your little finger nail. Ticks carry tick-bite-fever and prefer to attach themselves around your groin. For years we used to take them off by blowing out a match as soon as it was lit and using the hot head to burn

## REMEMBER:

Mosquitoes need still, preferably stagnant, water for their larvae to grow up in. Take away the water and you remove the mosquitoes from an area. The trouble is you have to get every bit of water so be vigilant about what you leave lying around camp. In one Far Eastern country which used to be overrun by mosquitoes they have been all but eradicated from the cities by locking up anyone who leaves still water around outside.

the creature. If the procedure was done correctly, and there was a knack to it, then the tick would release its grip and come away. The thing is if you just try to pull a tick off you it grips hard and the body comes away leaving the head inside your flesh. This will then become infected.

In recent times the French have invented a little gadget designed to remove ticks and issued it to their troops. You can buy one in any chemist in Europe now. It is so simple yet 100% effective. All you get is a little plastic handle with a 'V' notch at the business end. You slide the V between the skin and the head of the tick and twist. The tick comes away in one piece without any problem. Anyone can do it first time. I always carry one in Europe as you never know when a tick will strike – my dogs and I pick them up regularly.

## Worms

These come in many shapes and sizes but for the purposes of my simple explanation I shall split them into two types. There are skin worms and there are gut worms. Skin worms are laid as eggs, by insects, either directly onto your skin or onto plants which you then brush against to collect them. The eggs hatch and burrow under the skin as a worm. In parts of Africa the natives, seeing a worm under the skin as a white-ish line, make a cut at one end of the line and tease out the end of the worm. The end is then attached to a stick and the worm, often over a foot long, is withdrawn by winding it around the stick. No doubt a painful process.

Gut worms live in your intestine and share your food. A couple of hundred years ago fashionable ladies in Europe used to eat a few worm eggs to help keep their weight in check. So you will gather that the main effect, apart from infection and the gross-out factor, is that you lose weight as they induce nausea.

You will come across the eggs on grass and other vegetable matter. They are too small to notice but they are excreted there by other carriers of worms. The worm eggs stick to your hands or body and find their way to your mouth. Once inside they hatch, hook on the inside of your intenstine and grow happily. Take care eating fruit and other vegetable matter or touching livestock – foodstuffs and hands should be washed thoroughly to get the worm eggs off.

## Rats

In themselves, rats are nice little creatures and make excellent pets. If they get at your food however, they can be a problem. The issue is not what they eat, as they don't eat much, but what they spoil or, worse, what they infect without your knowledge. Rats crap on your food and their crap carries diseases such as typhoid and plague. Yes, people do still die in droves of plague every year. Rats also are prone to carry rabies. Though this is not easy to catch it is very nasty. Fortunately, you have to be bitten to catch rabies so watch your fingers. Less fortunately, a rat with rabies gets tetchy and prone to bite. You need to take the rabies anti-dote as soon as possible.

## Scorpions

I like scorpions. They are excellent little things. But they do have quite a powerful sting. For what it is worth, the bigger, black scorpions with big claws tend to have milder poison than the little pale ones with small claws. The thing is, even though a scorpion sting is not likely to kill a fit man it is extremely painful.

A friend of mine, a very tough 30-year-old paratrooper from the French Foreign Legion was stung in the middle of his back by a scorpion which crawled into his sleeping bag one night in Zululand. He was rolling over and over around the floor putting on a performance for some time. Everyone thought it was hilarious and we laughed until we cried. All of us except him of course.

Scorpions hunt little creatures like insects. They don't intentionally seek to bother people. But they do like a nice warm place to sleep during the day so beware of them getting into your sleeping bag or your boots. Make it a habit every morning to turn your boots upside down and bang them together to shake out any sleeping scorpions which were attracted by the warmth the previous evening. Presuming you get to take your boots off to sleep of course.

## Camel Spiders

Camel spiders – *Solifugae* – are a particular type of arachnid and not actually a spider at all. Neither do they eat camels. There are a great many horror stories told about them but though there are hundreds of types found in deserts all over

A camel spider eating a large cricket. Officially they can grow up to 8in in size but there have been sightings of far larger camel spiders by troops stationed in Iraq and Afghanistan. Their sting is nasty but not dangerous. (Corbis)

the world there is only one in India which is at all poisonous. They are quick and will nip hard if abused but otherwise they are harmless and just look like something from a sci-fi film. They are found in Afghanistan so be prepared to stumble across them if you are going on operations shortly.

## Other insects

As a general rule a brightly coloured insect is not trying to hide, it is saying, 'Watch out I am dangerous!' Avoid insects of this kind and as a general rule assume spiders to be poisonous. Many are but relatively few have venom strong enough to harm a human. But you never know which is which do you?

## Snakes

Outside the Arctic Circle I think it is only Ireland that doesn't have snakes, and this is courtesy of Saint Patrick. So wherever you go there will be snakes to a greater or lesser degree. The thing is, snakes, except for a couple of species out of the many thousands, are not aggressive towards humans. They feel the vibration of your approach, as they are deaf, and they clear off out of your way.

Crushing-type snakes like pythons can't harm you and the poisonous types would rather not meet you. They have their own lives to live which centre around catching rats and mice and so forth. Even in countries where there are many snakes you will probably never see one. The closest you will come, unless you are looking for one to eat, will probably be to see the tracks of a snake where it has crossed a dirt road or see the skin it has shed.

The only things you have to watch out for with snakes is surprising one in the latrine, in a thatched roof or in your sleeping bag. If cornered a lot of snakes will go for you and many will kill you. They are surprisingly fast. Take no chances. Assume the snake is poisonous unless you are an expert.

A friend of mine, the author Yves Debay from the Belgian Congo, came across a highly poisonous adder in his sleeping bag. Being of that frame of mind he killed it with his knife, skinned it and ate it raw. Snake tastes like chicken, better if you are hungry. Some Africans have a big issue with snakes, considering them to be taboo or evil spirits or some such thing. Yves was held in awe by many Africans. Some Europeans too come to think of it.

## Big critters

Of all the big cats, leopards eat by far the most people and even come into town to eat dogs left outside at night. They won't approach a bunch of men, and neither will lions. Tigers, however, will occasionally creep up behind and take a man from the back. Where there are tiger, the locals often wear a mask on the back of the head as they allegedly won't attack someone facing them.

Elephant are smart and don't look for trouble unless they are males and it is breeding season – when they go a bit crazy like some Airborne types I know. If you bump into them they will generally walk off after a bit of a song and dance. Hippos can be very nasty in the water as they protect their territory. But out of the water they are about the same as rhino – bad tempered but hassling you is too much trouble unless you really piss them off.

Water buffalo are no trouble but African plains buffalo are the most dangerous creature on earth. If you shoot one or two they will keep coming until they overwhelm you. If you climb a tree they will wait around the bottom until you fall out of it. Leave them well alone. Many years ago a civilian friend of mine was working in finance in East Africa. He was out on some sort of safari and a buffalo attacked him. It managed to hook one of its horns in his arse-hole and swing him about a bit like a balloon on a stick. Did him no good at all.

# FITNESS FOR BATTLE

Many people think, because there is such an emphasis on fitness in the training of soldiers, that fitness and athletic performance are all-important components of a soldier's efficiency.

The truth, as is so often the case, is a little more complicated than that.

## Keeping a clear head

As you will gather from elsewhere in this book, and personal experience when you get it, soldiering is mostly about endless walking carrying heavy kit. Perhaps less so with the increasing use of Armoured Personnel Carriers (APCs) for infantry but when you have to patrol in town or country, and especially as Special Forces, your soldiering has to be done on foot at least some of the time.

It doesn't take a great deal of fitness to do a four-hour patrol on a spring day but it does in the desert sun. And the thing is, any insurgent worth the name will know to hit you when you are on the last half hour of your patrol. Just like they try to hit you in the last days of a tour. If you are not fit by the end of a patrol your concentration will be ebbing and your reactions will be slow. This will get you or your mates killed.

So, the purpose of general fitness in a combat soldier is not just the ability to make the occasional mad dash to engage the enemy or the skirmish forward that happens in conventional warfare now and again, it is the ability to keep a clear head and rapid reaction time after hard work — ready for anything exciting that happens wherever you happen to be.

## Training

The last thing you want in a soldier is to be over weight. A fat soldier cannot do his job; firstly because he probably can't carry his body plus his kit the necessary miles but also because part way through the patrol he will become dizzy with exhaustion and switch off. This inability to stay alert and spot potential trouble makes him a liability to himself and to his mates.

The next worst thing is a bulked-up, muscle-bound soldier. Yes, really, despite what you see on the TV! This is because muscle is great for looking tough in movies and lifting heavy weights but it is a real drag on the heart when you have to run or walk long distances. You don't get any benefit from big biceps when you are running but they constitute extra weight to carry and, worst of all, your heart has to feed them with oxygenated blood which detracts from what is available to power your legs. What you want is to be slim and fast. A soldier should look like a marathon runner; and all long distance runners are thin.

## Test of willpower

Elite units and Special Forces often have a selection procedure which involves either extreme standards of fitness or, and this is slightly different, the requirement to cover great distances on foot, carrying weight in a pack.

## TOP TIP!

Training gets boring after a while so I suggest you play football, basketball or similar for fitness and do long marches several times a week for stamina. I personally used to box as an incentive to fitness. This is a great motivator as if you are not fit in the ring you get hit a lot.

Very high standards of fitness tend to be one of the factors which increase aggression in young men. This is just what is wanted in a force like the British Parachute Regiment or the US Airborne and the leadership has to put up with a certain amount of exuberance. Fighting in bars and so forth.

Covering great distances on foot day after day while carrying a heavy pack does require fitness but not extremely so. What it does call for in large amounts is will power – the mind-set that keeps a man going when he is tired and blistered. Through experience it has been found in the British Army, and the SAS in particular, that a man who will keep himself going when his body has had enough is the sort of man who, without the support of comrades or an officer to tell him what to do, will keep going and accomplish his mission.

To give you an idea of what is required, consecutive daily SAS selection marches are in the 20+ mile range, testing marches are 30 to 40 miles and we once did a march of 65 miles, over mountains at night and carrying 65lb packs in 14 hours and 20 minutes.

Tom doing the 'Fan Dance' carrying a GPMG: all British Special Forces soldiers climb Pen Y Fan as part of an endurance march. For decades the Fan Dance has been a crucial part of SAS selection. Part of the Brecon Beacons and the highest mountain in South Wales, Pen Y Fan feels higher than it actually is when you are carrying a 44lb Bergen and rifle. As recently as 2011 a recruit died just 700m from the finish line during the selection process.
(Photo courtesy Tom Blakey)

## Shin splints

Some soldiers are more prone to 'shin splints' or 'stress fractures' in the lower leg than others. What happens is that continual running over months or years, in boots particularly, and on hard surfaces such as concrete or asphalt causes a crack to appear in the shin bone by the constant shock of the heel hitting the ground.

I used to get this as a recurring problem and the medics experimented on me with heat induction to speed healing. Maybe there is something clever they can do nowadays but the best thing is not to get a splint in the first place. In the 'old days' we always had to run in boots. And there was no 'give' or spring in the heel. What you want is to run in trainers with cushioned heels and march in boots. Don't worry too much about shin splints — I had to run 12 miles and Tab 12 miles every day for years to get them. In any event, it's only pain.

## What to do?

To make your soldiering easier, keep your weight down and your fitness high. The best way is to make training interesting by playing aerobic team games such as football/soccer, basketball or similar. And keep away from the weights.

The Author — way too heavy for an infantry soldier. (Author's Collection)

# CHAPTER 5

# How to Avoid Getting Shot

In some respects you might consider this to be the most important section of this book.

The idea of following the occupation of a Special Forces soldier, indeed any soldier, is not to be a dead hero, survive awful wounds in the pursuit of your duty or even win a hatful of medals; it is to stay alive and healthy while completing your mission – be that keeping the peace or fighting a full-on war.

The trouble is, while you are completing your mission – be that protecting a food convoy or attacking an enemy base – the enemy is going to be trying his damned best to kill you just the same as you are trying to kill him. And the thing is, no one is totally bulletproof, so the best thing you can do to stay alive in combat is tilt the odds in your favour by fitness, skill and following good SOPs. Pretty much everything you learn in training and perform in the field is to stop you getting shot or blown up so there are many other things you can do too, but I think by listing some of the principles in this section it will help them stick in your mind, and maybe that will stop you getting shrapnel stuck in your butt. In principle, you can achieve the worthy aim of keeping your skin intact by one of three methods:

1. Shoot straight and get the other guy first – dead enemies are no threat.
2. Avoid being seen and therefore shot at, by your expert fieldcraft and camouflage. The enemy can only shoot what he can see.
3. Stop the bullets and shrapnel which are coming your way with some effective body armour. Skin is pretty tender stuff but fortunately there are some tough fashion statements available to the modern soldier.

# SHOOT FIRST AND SHOOT STRAIGHT

I am going to tread on some tender toes and upset people again here but my brief is to keep you alive – I don't care about happy. That's your mother's job. Winning a shooting match is the best way of not getting shot. The simple premise being that a dead enemy is not going to shoot you. The trouble with this concept is that the enemy has exactly the same idea.

And, contrary to what a lot of armchair warriors say, insurgents are often very brave soldiers. They are not generally stupid either – even if some of them have some strange ideas about how their God wants them to live. We kill ten or a hundred of them for one of us by winning shooting matches on a regular basis not through mindless heroics. Why do you think

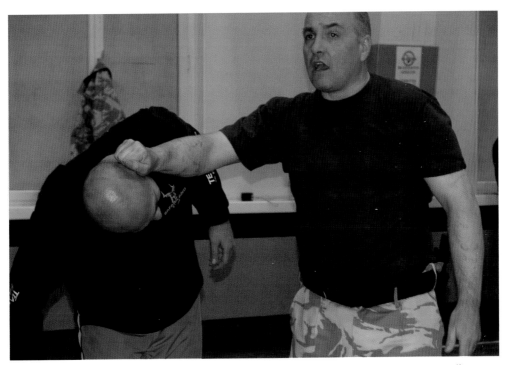

Whenever you possibly can, cheat to swing the odds in your favour. The Author demonstrates dirty tricks on a willing volunteer at the Military Academy, Vienna. (Author's Collection)

that is? I will tell you here and now that the reason we win more than we lose is that we have better organization and we have better training.

**Organization** wins conflicts because it means having leaders at every level who know what they are doing so they come up with the right plan at the right time and then organize its execution properly. That might be your section commander, your platoon lieutenant, your squadron major, your battalion colonel or the general who makes all the big plans.

**Training** means every soldier is trained to operate efficiently like a cog in a machine. Reacting to a certain threat in a particular, effective, way by means of carrying out SOPs. Every soldier must know his personal soldiering skills inside out and be able to carry out the plans he is involved in exactly and to the letter. Playing your part in the plan means knowing your own job and, to my mind, understanding your part in the bigger picture in case something goes wrong and you have to improvise.

Get this right and we will win with minimum casualties – you may even get to stay alive forever. Get it wrong and men on our side will certainly die. Now for most of a soldier's career

Intense training undertaken by units such as the US Navy SEALs ensure that they are crack shots no matter the terrain or circumstances. (US Navy SEALs and SWCC)

there is someone above him making the plans but I have given you an introduction to tactics in this book so you have some idea what is happening. What we really need to look at in this section is your personal soldiering skills so we will start with covering a bit about shooting that you might not have learned on the range and a bit about 'fire and movement' that you might have forgotten.

## Combat shooting skills

The simple fact is that killing the enemy and staying alive are much more a function of aggression and clever tactics than good shooting. The **safest** way to fight a war is to open fire at several hundred metres like you see so often in the sand pit. This way each side fires off loads of ammunition until one side runs low then that side clears off. Casualties are very few but everyone thinks they had a hot battle. Bullshit, they wasted an opportunity to defeat the enemy.

Do you realize that currently only one enemy soldier is killed for every 10,000 rounds fired by NATO and ISAF forces? What sort of shooting do you call that? Pretty rubbish shooting by many standards but it is all down to opening fire at too great a range and this happens principally due to poor tactics. Get the enemy close before you open fire or you will waste ammunition and take risks for nothing.

## My tally

In an effort to try to get you to take this seriously I will tell you something about my own efforts. Few soldiers talk about killing with the exception of snipers. But snipers are killing machines and see the enemy as scoring targets on a video game. Or they go mad.

I have killed more than 100 men in combat. Less than 10% of these have been more than 50m away when I pulled the trigger – though often in a firefight you are never **exactly** sure who made the kill. And less than 10% were within 5 or 10m. So nearly everyone I have shot has been between 10 and 50m away. How hard is it to shoot someone at 50m?

What does all this tell you? Several things I think:

1. To actually kill the enemy efficiently with a rifle it helps to get up close and personal
2. To get up close requires commitment on your part and gives you the edge
3. You don't have to be a very good shot to hit someone at less than 50m
4. Up close you really, **really** do not want a stoppage

So what is there to learn about shooting if you are never going to shoot at long range? Plenty...

The Author training Special Forces troops in the art of snap shooting somewhere in Africa. Stock tight into shoulder, look over barrel. Balance weight between feet equally. Grip fore-stock well forward with vertical forearm. Pull weapon tight into shoulder with both hands. Raise rear elbow level to seat stock of weapon. Rise barrell onto target. Squeeze trigger – do not pull. (Author's Collection)

## Maintenance

Keeping your rifle working is probably as important as anything. If your rifle doesn't fire when you pull the trigger then you are certainly no use to your team and you will probably die if someone comes for you. The aim of maintenance on any weapon is cleaning away the carbon build-up from the working parts and preventing dirt build-up and rust on the others. You clean away carbon with a wire brush, chemicals and a scraper then you apply oil as instructed for the particular weapon.

> **REMEMBER:**
> When you strip your weapon in the camp lay out the parts neatly but when you do it in the field place them in an upturned hat or similar to avoid loss.

You should practice stripping your weapon for cleaning until you can do it with a blindfold on – seriously. It is an excellent way of getting a soldier familiar with his weapon. Plus it is useful for when it is dark. If you allow carbon from the propellant burnt with each round fired to build-up then the rifle will start to jam as the fresh round will not feed into the chamber correctly, the breech will not close and lock to allow firing or the gas pressure will not be sufficient to drag the sticking empty cartridge out of the chamber.

Your weapon needs to be protected from the main enemies of all mechanical toys: rust and sand. The more finely crafted the weapon the more it suffers from a few grains of sand in the working parts. Most Western weapons are made to such fine tolerances that a small amount of sand brings everything to a halt. The Russian and Chinese Kalashnikovs, on the other hand, are made so loose that the working parts rattle when you shake them. A surprising amount of sand or mud will entirely fail to stop them working. There is a lot of sense in that. But what do you do with the Swiss watch of a rifle with which you have been issued? Consider this:

- **When it is wet** then there is no dust in the air so your rifle's main enemy is rust – given the one you have still has steel in it. Therefore, after your daily cleaning, wipe plenty of oil on the working parts and other areas at risk of rusting.
- **When the weather has been dry for a while** then dust and sand are the danger so when you have cleaned your rifle apply oil very sparingly and wipe it off. This is sufficient to protect from rust yet not enough to cause sand to stick to the working parts.

## Stoppage drills

Given your rifle is clean, and your ammunition loaded properly, it will rarely jam. Aside from carbon build-up the main causes of weapons jamming are a bad feed from the magazine, damaged or dirty ammunition and sand or dirt blowing into the working parts. Let's look at these in turn:

The easiest way to get a bad feed is by allowing the rounds to work forward in the magazine so that when they are picked up by the advancing working parts they approach the breech at the wrong angle and jam across the face. Make a habit of tapping the back of your magazines against your weapon before you fit them. This tap will seat the rounds where they are supposed to be – at the back of the magazine.

If one of your rounds has got dirt stuck to it or the bullet has come loose and is 'bent' relative to the cartridge then this will cause a jam similar to the above. Check your ammunition when you load your magazines – by touch if you have to do it in the dark.

The standard stoppage drill for most weapons is 'Cock, Hook and Look' which is pretty self explanatory: pull the working parts right back and then engage a little lever which holds them there – if supplied on your weapon. Otherwise hold the parts back and look inside. If there is a round jammed by the working parts, and this is nearly always the case, pulling them back often clears it and you can turn the weapon over so it drops out. Let the parts move forward to load the next round and continue shooting.

In combat, if your rifle jams, cock it again and it will probably load so you can continue. If that doesn't work, take off the magazine, check the rounds are seated to the rear, refit and let the parts move forward to load the next round. The cause might have been dirt in the working parts. It is surprising how quickly you can do the first stoppage drill when someone is shooting at you and 99.9% of the time it is all you will need to start shooting again.

## Weapon handling

Before you come anywhere near a shooting match you must have had at least basic training in handling your rifle and shooting on a one-way range. What I suggest is that you should be able to shoot with reasonable accuracy from lying, kneeling and standing positions with either hand. Why? Because if you are right-handed and looking around a corner which is up against your right shoulder you have to expose your body to shoot. Likewise if you are left-handed and looking around a corner to your left. Train hard, fight easy.

Very often, when you come to shoot someone, you are in a bit of a hurry to get the shot in first, probably due to nerves or excitement if you are new to the game. The problem is that your relatively heavy rifle barrel is going to be swinging quickly towards the target and then past it. If you don't fire at exactly the right moment you will miss. How do you get around that?

There are two things you can do to improve your snap shooting skills. The first is to swing low onto the target then lift the sights onto the target. This results in the final movement being up rather than across and is much more accurate as I mentioned with a pistol earlier. The second trick is to hold for a count of one before and after firing. This stops you snatching the aim away to shoot somewhere else before you have pulled the trigger. Believe me, it is a natural reflex when you are rushing and really does happen.

Whatever position you are firing from you should pull the weapon hard into your shoulder as this gives a firm, repeatable-firing position and leads to greater accuracy on both the one-way and the two-way ranges.

Raise the elbow of the shoulder in contact with the butt of your rifle until the upper arm is level. This feels strange but also gives a stronger repeatable firing position.

The forearm supporting the fore-stock of your rifle should be as close to vertical as possible and gripping forward of the point of balance on the weapon. The reason for the forearm being vertical is that there is a tendency for the weapon to rock or hinge very slightly on the elbow below it and keeping the forearm vertical both minimizes this and causes the grouping of rounds to centre on the middle of the target. The reason for gripping well forward is that is reduces the swing of the barrel as you are sighting.

Place the second pad of your index finger, between the first and second joints, against the trigger – not the pad at the end of the finger. Do not pull back to work the trigger but squeeze your finger smoothly towards your thumb. One of the greatest causes of inaccuracy in shooting is jerking the trigger and pulling the weapon off aim at the last moment.

## Ammunition conservation

There is one other thing which can leave you of no use to your friends and at the mercy of your enemies. You can run out of ammunition. This is worse than embarrassing. Absolutely everyone gets a little exited the first time they are shot at and this makes them tend to fire off more rounds in reply than is strictly necessary. When you have been shot at a few times you will think nothing off it – and by that I mean shot at close enough to hear the 'crack' of

## TOP TIP!

### Aiming at your target

Always aim for the middle of the observed target, sometimes called 'centre of mass'. This allows for you to make the greatest error and still hit your man. Any hit is a good hit as it will certainly prove an inconvenience to your target and allow you to approach and finish the job.

A standing man is a taller target than he is wide. Obviously. Therefore you have more leeway up and down where you will still hit him than you do left and right. In other words, you have to be more accurate left and right than up and down. Got that? OK.

the supersonic round as it passes over your head followed by the 'thump' when the sound of the propellant firing reaches you.

The next stage in your initiation is when you have returned fire a few times at likely sources of irritation and after that when you have aimed shots at a few breathing targets. All these experiences allow you to calm down while you are being shot at and you will fire back more accurately. When you have got to this stage you are something of use on the battlefield.

So you are calm and collected in the fire-fight but what should you be doing? Well your weapon fires around 10 rounds a second on automatic so this way your ammunition doesn't last long. As a general rule I suggest you fire single rounds at targets, short burst when ordered to provide suppressing fire onto an area and only blaze away on automatic when you are caught in an ambush.

There are one or two weapons which actually have see-through magazines so you can see how many rounds are left. I tested some in Germany and Austria recently, and thought what a great idea. I hope to see this spread as it is quite difficult to count rounds when using automatic fire and this can lead to a penetrating click at the firing pin strikes home just when someone carrying an AK47 is approaching you with evil intent.

## Individual movement techniques (IMT)

I was taught this technique a lifetime ago and I thought it a little questionable even then. Today I think it is plain stupid and wasteful of men but I will explain the concept and leave you to make your own decision as to how useful it is.

Firstly, have you ever heard that phrase before? If not, you are not alone. This section is about what I learned as 'Fire and Movement' or the 'Buddy Buddy System'. The thing is, just about every army in the world has it and uses a different name for the same thing. The idea is that when you are out on patrol and you come under fire there needs to be an SOP for you to get on with for a few moments while the unit commander figures out what to do. Individual Movement Techniques is that SOP.

Suppose you come under fire from the flank while you are walking in single file. What do you do? You turn to face the enemy and work in pairs to advance on him. One man takes up a firing position and lays down suppressing fire while the other moves forward a few yards or as the ground allows. Then he lays down

Two Pathan tribesmen examine a painted Lee-Enfield .303 rifle at the village of Darra Adam Khel, 26 August 1980. The old bolt-action rifles are still incredibly accurate at long ranges. (Corbis)

suppressing fire and the first man moves past him. By this means the unit advances on the enemy steadily until he either runs or is destroyed.

In combat, this is supposed to allow the first few moments of the engagement to occur almost automatically and gives the soldiers a way to respond appropriately and predictably while the unit commander evaluates the situation prior to issuing orders. I think this is a load of bollocks. All it does is get men killed. The correct answer to coming under effective fire is to hit the ground or run for the nearest cover as seems most useful – shooting as you go. Then you try to get some rounds off at the enemy to suppress his fire while the unit leader decides what you are going to do. This might be fire and movement towards the enemy or it might not. More likely it will be suppressing fire while you call in air support or while you set up a fire group and an assault group to advance on the enemy when he has his head down.

## Bullshit...

Let me just get something else out of the way here. If you look at some of the thousands of soldiers' groups on Facebook, or magazines relating to soldiering, you will see all sorts of fancy guns and gadgets to make your gun work better. There are special grips, special aiming sights, special light fittings and on and on. This is almost all entirely bullshit.

Gadgets are to reassure a timid soldier who is not confident of his, or his team's, ability. I want you to be so good at your job you have confidence in yourself and your mates not your hardware. Then if the kit goes missing or breaks down you still have the important part of the operation – yourself. Do you find that hard to believe? Consider this:

### Super soldiers?

Imagine a dozen SEALs or SAS operators armed only with bolt-action rifles from World War II. They are loose in a forest and set against a dozen infantry soldiers armed with every gadget and new bit of kit they can carry. Who would you bet on? The Special Forces I would imagine.

Why? Maybe you picked the SF because you think they are supermen. If so, you are right for the wrong reason. The reason the SF would win with no or relatively few casualties is because of a couple of simple factors:

**A good plan:** They are thoroughly trained and have experience so they are calm and collected. And so they can put a good plan together using the right tactics for the situation.

**Commitment:** When they have settled on a plan they **will** carry it out with no hesitation or lack of commitment. In their minds there is no more thought of losing than a butcher might expect a piece of beef to get up and walk away. And, of course, they will never give up.

### My Plan

I was asked by my editor what I would do in this situation. She thinks people might want to know how it should be done. So at risk of going off the subject...

**Training first:** I would get my team very comfortable with the old rifles and get them 'shot in' so the iron sights are set correctly for each man. A decent shot will maintain a 12in group at 300 yards or better and bringing that group onto centre of mass will make him deadly.

**Remember KISS?** (Keep It Simple Stupid): We would hide for about three weeks until the infantry soldiers got tired and cold and sloppy from not seeing us. Most likely they would be patrolling every day and getting wet and tired then doing stags at night. I would use one or two men to watch their movements and see where they went each day while the others made themselves comfortable as far away as possible. Then, when they let their guard down, I would launch one ambush from 200–300 yards and expect to kill half of them. Then withdraw. Why? Because the old bolt-action rifles have the advantage of greater accuracy and range over the modern infantry rifle. In response I would expect the survivors to take themselves and their wounded to a position they could defend. It would be interesting to see how well they dug in. Again, I would give them two or three weeks to think we had left, then I would hit them at dawn from the greatest distance we could get a clear shot. This might leave one or two alive so let them stew for a week or two and repeat. Job done.

# FIELD-CRAFT AND CAMOUFLAGE

Field-craft and camouflage is how you avoid being seen, and therefore shot at, while you are moving through country or laying quietly waiting to give someone a surprise. First of all you need to make yourself as invisible as is practical in the circumstances: according to your operational needs this might mean wearing a 'ghilly suit', named after a person who creeps up on deer for a living in Scotland, or just making sure you have no bright or shiny articles of equipment showing.

Hopefully you have been trained in the basics of field-craft and camouflage as it may relate to your area of operation. What I shall do here is cover some of the basics in case you weren't listening to the instructor

A US Navy SEAL sniper illustrating his skill at concealing himself although the use of black gloves makes this a less than perfect example of camouflage. (US Navy SEALs and SWCC)

Lying in undergrowth, a camouflaged British infantry soldier looking down the telescopic sight of the new British-made Long Range L115A3 sniper rifle on Salisbury Plain, Warminster, England. Black boots and black sights stand out well don't they? (Corbis)

or have forgotten. Then I will try to give you a few tips that might just give you an 'edge' – enough of an advantage to keep you on the winning – or at least surviving – side.

When I was a young lad of 13, some 40 years ago, I was taught in the Army Cadets that concealment meant giving due consideration to: Shape, Shine, Shadow, Silhouette and Spacing. This has been slightly improved over the years with a new mnemonic – if you can remember them all:

- **Shape**
- **Sudden Movement**
- **Sound**
- **Concealment**
- **Shine**
- **Silhouette**
- **Spacing**
- **Shadow**
- **Smell**
- **Sign**

## Shape

Our eyes – or brains – are designed to pick up on familiar shapes. The shape of a black rifle or a body lying on the floor will catch our eye much more if they are all the same colour and we can see all the shape which makes up the object. The two main ways to break up a shape are to colour different areas differently, like a camouflage suit, or to hide part of the shape from sight – keeping half of you behind a bush for instance.

A human face catches the eye too so when you are watching someone it is often a good idea to drape a veil across your face. This is a sort of string-mesh cloth in a square a couple of feet across. From the inside this hardly obstructs your vision at all but from the outside is blends in with cover extremely well.

## REMEMBER:

**Rank insignia:** Don't show rank insignia – 'Stripes' and 'Pips' – on active service. Any sniper worth his salt will hit the leader first followed by the radio operator followed by the gunner. On ops the team will know who is who and you don't want the opposition to know. Also, when giving orders don't point or wave when there may be a sniper about as it soon becomes obvious who the boss is.

### Shine

There are very few bits of shiny kit still in everyday use. Brass buckles are a thing of the past. But polished boots, metal equipment and worst of all your sweaty face can still shine alarmingly in random light when it is dull or dark. Wear brown boots if possible, camouflage

If you want to live forever, don't make a silhouette. (Barry Lloyd © UK Crown Copyright, 2009, MOD)

your rifle and make sure you apply cam-cream to your face. Even if you're really dark skinned you still want to use black or green camouflage cream to dull your skin. We all get sweaty and shine by reflection at night.

With regards to the use of light be very, very careful when smoking, lighting cigarettes and reading maps at night. It's a good idea to watch someone doing all of the above in darkness to see how good a target you make. When you strike a light or shine a light on a map you light up all the faces looking at it and make a perfect aiming point for a sniper. Like so many things, this is where carelessness gets you killed.

**Muzzle flash:** Take notice the next time you do a night-firing exercise, or go on the range at night to see how your muzzle flash lights up the whole area around you. In a firefight at night you should move after each burst – wriggle backwards, roll sideways, then forwards to shoot again. Or at least duck into your hole for a few moments if you cannot move. If you don't do this all the opposition might be shooting at you the next time you fire after a lull. The first to fire after a break in shooting very often gets everyone else firing at them so don't let them have you in their sights and be waiting for you.

### Shadow

There are two things to know about shadow: one to avoid and one to take advantage of.

**Avoid:** When you are in open ground and the sun is bright you will make a shadow, obviously. Given you, your hut or your vehicle are cammed up and keeping still you might think you are fairly invisible. Not so. When the sun is above you your shadow is small but as the sun goes down your shadow can become much bigger than yourself and, worst of all, it is black on the ground. Particularly when viewed from the air or from a hill looking down, the shadow is often the first thing you see of a person, vehicle or building if they are the same colour as their surroundings.

What can you do? With your body make sure there is higher cover than yourself next

## REMEMBER:

Remember to consciously aim lower at night as the instinct is to shoot high if you don't have proper night sights. It is amazing how people shoot high at night when firing at each other's muzzle flashes. If you can aim down you will hit ten times as many of the enemy.

to you. When you camouflage vehicles and buildings make sure you stretch the netting out in the direction of the shadow so the enemy see the net and not the shadow.

**Take advantage:** When you are viewing or shooting from a window, dark hollow or shaded area under a tree make sure you sit well back from the sunlit area. This will make you far less visible, the enemy will be less able to judge where you were after your muzzle flash and less able to hit you if you are in a building because you will have more protection from the window surround.

You can hardly see the crosses but look how visible the shadows of the crosses are. This illustrates my point. (i-Stock)

### Sudden movement

I think men evolved to see moving prey when hunting. True or not, it is a fact that anything which moves catches your eye. Even when well cammed a walking man catches your eye far more than a static individual. When you have to move in view of the enemy then either move very, very slowly, for instance if you are coming up to the aim, or move quickly and get it over with.

**Hard targeting:** When you have to walk along a street, and you don't know if there is a sniper watching, you need to 'Hard Target'. What this means is you avoid giving a steadily moving target for the sniper to set his lead on. You will recall that if you can get a moving target to move directly towards or away from you then it might as well be stationery for the ease of hitting it. If a moving target is going left to right steadily then you just have to aim a certain distance ahead of it.

If a target is hard-targeting, that is a couple of steps forward, then a step sideways, then a step back, then forward again all at random it is extremely difficult to hit from any distance.

> ## TOP TIP!
>
> ### Flares
> If a flare lights you up hit the ground and freeze. Wriggling around for a better position will get you noticed.

We were first taught this for patrolling the streets of Northern Ireland and it is both hard work and mildly embarrassing to do but you can make yourself a very difficult target — if you can stand looking a fool. Put in the effort to hard target and you will not get hit — the sniper will hit the lazy man who cannot be bothered.

### Silhouette

Look through cover — such as a bush, rather than around it. This way you don't attract attention by breaking up the skyline with a familiar and moving silhouette. If you can't look through cover then look around the side of cover rather than over the top. Looking over the top of something is the best way to get spotted.

When on a cross-country march try to avoid walking along ridges. Anyone lower than yourself can see you as plain as a pike staff as a black silhouette against the sky. Deadly.

### Smell

I think Western men use their sense of smell a lot less than women. Have you ever seen a man smell his food like women often do? There is probably something there from our hunting origins where men would be killing meat and women would be testing fruit and berries by smell. Although having a poor sense of smell can be a big help when living in cramped conditions with a lot of smelly marines.

The opposition, however, depending on where you are in the world, may have a very well developed sense of smell and use this to find you in heavy cover. Particularly in the jungle or heavy bush where you will often be fighting almost at touching distance and a man smelling of tobacco or after-shave or garlic etc may give a warning to your enemy and either spoil the ambush or get you all killed.

## Sound

Depending where you are there may be animals, birds, insects and road traffic all making their particular noise. What catches the listener's attention is a noise out of place. Especially a repeated noise out of place. You can get away with one gunshot in a bar or city street. People will just stop and wonder what they heard then convince themselves it was a car backfiring. Two shots and there will be screaming and everything that goes with it.

In a normal combat situation you need to be quiet at key times – like in ambush positions – so you need to keep in mind what causes the problems. As a rule the metallic clang of a mess tin or rifle working parts sliding shut is a definite sign of soldiers. Low talk carries but it could be anyone talking. The 'squelch' on a radio – if yours do that – is a dead give away so better to switch off than risk that. If you have to talk then try to do so in a trench or hollow to hide the sound. If that is not possible then wait for some local noise and use it as cover for your own. Before setting off on patrol have all the men jump up and down to check they don't rattle.

## Spacing

Spacing catches the eye or saves lives depending where the spaces are.

**Good spacing:** When you are on patrol make sure you keep 5 yards between each man. This gives you two benefits. If a mortar comes down, someone treads on a mine or an ambush is sprung then it reduces how many men are killed. It may even cause an ambush not to be sprung if the enemy think enough will survive to turn on them.

**Bad spacing:** Anything which is spaced out evenly catches the eye. A bunch of fairly well camouflaged heads evenly spread along a hedge line will catch your eye as something artificial and make you look closer. Spread more randomly and it will not be as obvious. A small thing but small things add up.

## Sign

When you leave an overnight camp or an ambush position be sure to take with you as much as possible of your rubbish. What you leave behind can often tell the enemy a surprising amount about you. The most obviously useful fact is how many men you have. Even a good tracker cannot be totally accurate numbering a group of men walking in each other's boot prints. But if you leave a few cigarette butts at each sleeping or firing position, or you leave wrappers you can often give away your numbers. This might not matter or it might kill you all – don't take the risk.

Probably the most unpleasant result is from you leaving behind something personal which can give away an address or phone number belonging to your loved ones. Your nearest and dearest are quite likely to pass on information like this in letter so don't carry letters – or mobile phones with a call record or list of numbers. And that is all of them. If you leave a phone behind, at best the bad guys might get in touch with your loved ones and tell them what they are going to do to you. At worst they may pay them a visit. Take care of yourself and your family by not being stupid.

## Concealment

This sounds so obvious but I have to say it: the bad guys can only shoot at you when they can see you. When you travel try to stick to tree-cover, stream beds, dead ground or whatever. When you stop for a smoke, for a brew or the night, try to get out of sight.

This next titbit should be tattooed across your forehead: when you are in a fire-fight show as little of yourself as you can. Do not climb up on the trench rim to get a better shot. Do not stand up hosing fire at the enemy. It looks great on war films for kids but it will get you killed.  Make yourself as small a target as you possibly can and release accurate, aimed shots at the enemy.

## Crowds and mobs

I have said before that I feel very uncomfortable when faced with a mob. You probably will too the first time. Don't let crowds get too close or someone will stick a knife in you and/or take your weapon. Point your rifle at the crowd and use the muzzle to fend people off before you open fire. Keep your rifle or your pistol attached to your body by strong cord or wire. Keep the wire short to make it difficult to turn them on you. If a crowd turns ugly bring the team close together as then defence against the mob becomes more important than against a sniper. I have limited sympathy for guys who get lynched.

If a crowd gathers for no obvious reason it may be to distract attention away from the movement of weapons or explosives. Or it might be to draw you together to make a target. Do as the situation demands and keep your wits about you. Try to think of what the terrorist might be wanting you to do and then do something else. Keep watching the upstairs windows and flat roofs in a town as that is where the snipers will be. People rarely look above eye level in normal life so you need to get into the habit.

Don't go for your pistol or you may lose it. Don't try to use your rifle butt. If it gets to look like a lynching toss a grenade over their heads so it lands a few yards away and duck. The crowd will shield you from the blast and then be stunned for a moment – then you come out shooting. Argue the toss with the lawyers afterwards.

## Movers

If you burst into a room containing terrorists/hostage-takers AND hostages, shoot the movers first, then the men who are armed. This because anyone moving is over the shock and therefore a danger. The armed men may be frozen but the fact they are armed means they are not hostages.

My own opinion is that you should never enter a room in this way – though it is taught to special police units and Special Forces the world over. Throw a nice grenade in first then walk in and finish off the bad guys. The hostages will patch up. But your commanders will no doubt disagree with me and ultimately you have to follow orders.

About the best you can do with a standing search: Iraqi civilians are searched on the outskirts of the city of Basra in Iraq in 2003. (Corbis)

## Body Search

I'm sure your sergeant has told you but when you search a person — obviously living or apparently dead — it is a two-man job. It is better to make them lie down with arms and legs spread, but obviously this doesn't make any friends amongst the locals. One man stands at a safe distance from the head end and covers the searchee while their partner does the searching. If the body appears to be dead already then beware a grenade or other explosive hidden under the body. If the searcher rolls the body towards themselves first then a grenade may become apparent and the body will offer some protection.

With a live searchee, feel the crutch area carefully because if someone is going to hide something 9 out of 10 times that is where it will be owing to the embarrassment most men feel for 'touching up' another man. Then check the small of the back at waist height as this is another favourite area which allows a weapon to be hidden, is often missed in a quick 'pat down' and allows for a quick draw. I have used it myself many times.

Don't be shy about searching women either. I have known women carry explosives in hollowed out dead babies strapped to their backs. Generally, to search women you need women soldiers or military police. I'm old fashioned in many ways and wish it were not necessary to have women in a theatre of war but they are vital when dealing with civilians — not that women don't make frighteningly good soldiers when they want to.

## Car search

If you stop a car get the occupants to open the bonnet (hood) and boot (trunk) while you stand well, well back. These compartments may be set to explode automatically on opening.

Likewise, when you come across a car parked suspiciously as if it might be a bomb waiting to go off at an intersection or significant building, **do not approach it yourself**. This is what bomb disposal guys are paid for. Keep away and let the bloody thing blow up. You are worth more than a car and a few windows. Of course, this strategy has caused me to be the butt of many jokes and take some criticism but, you will observe, I am still here, typing with both hands and I went for a cycle ride this morning.

### REMEMBER:

A sensible enemy will hit your unit when you have just started your tour and don't know the ropes or when you are about to go home and have let your guard down. So don't.

# ARMOUR FOR SOLDIERS

Combat injuries almost all come from being hit by pieces of metal flying though the air: bullets of various kinds, shrapnel from grenades and mortars, splinters from heavy cannon shells, bombs and mines. What it amounts to is that either a bullet of some kind has been blown down a tube at you or a piece of metal surrounding some explosive has been turned into a cloud of projectiles and driven in your direction by the blast.

A World War II-era British infantry soldier's helmet. (i-Stock)

Occasionally the projectile is glass from a window or gravel from the ground but the effect is always the same: your skin is punctured so you bleed, the projectile enters your body and damages a vital organ or bits are torn off you. Either way you are damaged and potentially killed.

Human skin is pretty tender stuff so if you can't entirely avoid being shot at or blown up you will need some protection. A sort of thicker skin. The trouble is that protection is both heavy and restrictive to your movement so it might be an idea to protect just the most critical areas of your body – the areas where, if that bit is damaged, you are likely to die. Pretty obviously, your body, or more specifically your torso, is more essential than your arms and legs as it houses all your vital organs without which you would curl your toes up in short order. More vital even than your body is your head where even a single splinter or heavy tap can leave you unconscious or dead.

## Infantry helmets

The solution to protecting your head has been around forever: a helmet of some kind. But what kind? In the modern era helmets were first issued to infantry in time for World War I. Artillery exploding as air burst over the troops' heads was one of the biggest killers followed by a rifle shot to the head with a high velocity, heavy bullet at long range when anyone looked over the top of their trench.

Tom models the latest British Army GS Mk7 Combat Helmet behind the GMG. This helmet is light, has good mountings for sights and the rear does not catch on armour and push the front down over your eyes. (Photo courtesy Tom Blakey)

It was soon discovered that no helmet light enough to wear could actually stop a bullet as a World War I rifle bullet would go through both sides of a steel helmet at 1,000 yards. The best compromise was to design helmets with broad brims so that most shrapnel coming from above would be stopped from damaging the head and periscopes were devised to look over the tops of trenches. This solution continued through World War II and after until wars became counter-insurgency operations. The significance of this is that a guerrilla army tends not to have effective artillery and therefore the counter-insurgency troops would be wearing a heavy helmet for nothing.

Very many troops have operated against insurgents in the Middle East and elsewhere without helmets right up to Northern Ireland and the present day. As society has softened its attitude to rioters and troops have become expected to tolerate a shower of bricks, rather than putting them down with fire, the helmet has appeared again in a form similar to an open face motorcycle helmet and made of fibre glass so as to protect the soldier from sticks, stones and other hand-thrown missiles. Today this type of helmet is still deployed around the world when troops have to contend with rioters but are not allowed to shoot them.

On the modern battlefield, which is primarily counter-insurgency remember, artillery remains a limited threat to our side but the enemy have the opportunity to use both hand-thrown missiles and rifles. Fortunately, technology has advanced to such a degree that the latest helmets, of the motorcycle style, are now able to defend to some extent against small-arms fire. Certainly a glancing blow from a Kalashnikov rifle at extended range will not penetrate the latest designs. That is quite reassuring is it not?

## Infantry body armour

From soon after Napoleonic times to well after World War II no body armour of any significance was worn by the military except for aircrew – to protect them from shrapnel produced by anti-aircraft shells bursting around their plane – and tank crews – to protect them from the shrapnel that scabs off the inside of an armoured fighting vehicle when it is struck hard from the outside.

You will notice that both these forms of armour were for protection from shrapnel. It has usually been shrapnel that killed most servicemen and in any event, rifles were so powerful that armour heavy enough to stop their bullets was not an option. Then in the 1960s the flak jacket started to make an appearance in general issue to troops. A sort of thick waistcoat of some sort of plastic fibre which was light but would stop most shrapnel and pistol or sub-machine gun bullets.

Studies had convinced the 'top brass' that by far the greatest number of casualties to all servicemen in World War II were caused not by bullets but by shrapnel so as per usual, the generals were equipping their men to fight the previous war. Some would argue that as shrapnel and low velocity rounds were not a real problem in recent conflicts the main effect of the flak jacket was to improve morale.

Flak jackets have continued to be issued to Western troops since the 1960s for use in riot and counter-insurgency operations. Over the years they have improved in design to become significantly more effective in stopping nasties as well as lighter and easier to work in.

## Infantry armour for the present day

Today the question is still where do you draw the line in the balance between weight and protection? One limit to the effectiveness of an infantry soldier is how much supplies, ammunition and explosive he can carry on a patrol. Already British soldiers are carrying up to 150lb of equipment under the Afghan sun so how much armour can they carry when every added pound of armour is a pound less ammunition or stores?

First of all we will look at the main threats:

Artillery is not used against us as air burst in counter-insurgency warfare but there is shrapnel from bombs, mines and rocket propelled anti-tank grenades. Kevlar helmets and standard body armour will stop all this to the head and torso. Fine so far.

But what about rifle fire? The main small arms threat comes from the Kalashnikov rifle firing a 7.62mm x 39mm round and the PKM type machine gun firing a 7.62mm x 54mm round. The machine gun round is much more powerful because it fires the same bullet driven with more propellant and therefore travelling a lot faster.

There is still no wearable helmet which will stop a rifle bullet head-on at close range. Though the latest helmets do deflect a glancing blow – which is comforting. The latest body armour, however, can stop rifle fire to the chest. How does it do that? The Americans have developed a suit of armour for infantry called 'Interceptor Body Armor'. The British have something similar called the 'Osprey Mark 4 Body Armour'. I find it strangely comforting that they are both very similar in theory and operation. That suggests to me that different, smart people approaching the same problem in their respective ways came to the same conclusions and therefore these may be right.

To keep it simple, both types of armour are primarily waistcoats of soft armour which will stop 9mm rounds and shrapnel. They both have pockets in the front and back chest area to insert stiff armour of various types. This armour **will stop rifle bullets**. Using this armour pack you have to make the decision as to when to put your

Body armour in action in Afghanistan. (Barry Lloyd © UK Crown Copyright, 2010, MOD)

plates in and carry the extra weight. My advice is to wear a shrapnel-proof helmet and shrapnel-proof soft armour at all times you are at risk from enemy action. Use your discretion as to when is the time to insert the armour plate. You might want to think about this option if you are driving around rather than walking, directly attacking the enemy or entering a building to rescue hostages etc.

## Armour for covert operations

If you happen to work as a Spook or in plain clothes as an Special Forces operative for whatever reason then the rules change because the threat changes. I am talking now about Intelligence work in exotic countries such as Russia, the Balkans, parts of the Middle East or West Africa.

**The threat:** The chances are you will be wearing light-weight clothing and need to keep both armour and weapons out of sight. Unless you come up against me you are unlikely to be attacked with a hand grenade. By far the most likely threat to an agent comes from a pistol. Now, as you may know, the mechanics of a pistol oblige it to fire a relatively heavy bullet quite slowly compared to a rifle. The bullet tends to be soft so as to do the most damage to a body but it has far less kinetic energy and penetration than a rifle round. The most common calibres you will come up against in the modern world range between the 9mm automatic and the .45in automatic.

**The solution:** You will most likely have heard of the expression 'bullet proof vest'. This term is usually used to describe a light-weight piece of body armour worn hidden under clothing. It covers a similar area to a T-shirt but is rather thicker. And warmer. There are various types and materials but in principle the bullet proof vest is what is called 'Soft Armour'. This term denotes armour which 'gives' under the blow of the projectile and will allow severe bruising to the wearer while preventing penetration and puncture damage to the protected areas. Do be aware that after sustaining an impact wearing soft armour you may not be able to stand immediately. The main thing is, you will be able to stand later. Often there is a facility to insert extra armour plates into the front and back of these vests.

Even without plates this soft body armour will stop all normal pistol bullets. The problem comes when someone has loaded up with some of the latest armour piercing pistol rounds.

Then your armour will not work except to hold your guts in. A friend of mine from the southern states of the USA and who went by the name of 'Rebel' spent some time as a mercenary soldier and other times as a bass player. He wore soft body armour habitually – with and without plates. You might wonder how good a guitar player he was. Only joking Reb...

# SURVIVING AIR TRANSPORT

As a soldier your generals are likely to move you all over the chessboard using air transport. There isn't a great deal you can do to protect yourself while you are flying – firing from an open chopper is only effective at low altitude – but that isn't what I'm going to talk about here. What I want to look at here is how you protect yourself and your aircraft when it is most vulnerable – as it lands and takes off.

An Osprey leaving the scene with the back door down and a gunner on watch just visible.
(Photo courtesy Tom Blakey)

We all know that aircraft are prone to damage from ground fire. Typically, fighter and ground-attack aircraft fly too fast and are too heavily armoured for anything but a specialist anti-aircraft missile to pose any threat. But that isn't what you are going to be flying in is it?

You are going to be moved about in large, ungainly, vulnerable fixed-wing aircraft like the good old Hercules or choppers of one kind or another. You may even get a go in the Osprey – issued to the US Marines – which can change between being a helicopter and a fixed-wing aircraft.

Apart from being fairly uncomfortable for the passengers and crew, the shooting down of an aircraft is a major propaganda victory for the terrorist. There may only be a couple of people killed in a fixed-wing crash or wounded in a chopper incident but the whole idea of the terrorist being able to bring down an aircraft has a certain newsworthy 'air' to it which gets the video clips shown on the news all over the world.

Perhaps it is the high technology they appear to be overcoming or perhaps it's the immense cost of aircraft they are destroying. Perhaps it is playing on the fear most people have deep down of the civilian aircraft they are flying in crashing. Whatever the reason, the terrorist will bring down an aircraft if he possibly can. You want to stop this happening if you can; and if he succeeds you don't want to be in it.

## Fixed-wing aircraft

A fixed-wing aircraft is most at risk while it is close to the ground and moving comparatively slowly. As they almost always take off from and land at airfields this is where the terrorist threat mostly lies. A fixed-wing aircraft is generally vulnerable to small-arms fire. Specifically the crew, the engines and the fuel tanks. As an accurate shot is difficult, even at close range, the terrorist will use machine guns, RPG rocket launchers or short range, shoulder-launched, anti-aircraft missiles in the hope of doing sufficient damage to bring down the aircraft. Obviously, when a fixed-wing aircraft comes down it completes the job of its own destruction.

Aircraft fuel tanks are not as vulnerable to small-arms fire as you might think. All tanks are fitted with a self-sealing system – which I won't go into for security reasons – but which will prevent fuel loss or fire risk from small-arms fire. The point here is that the range at which a light machine gun can hope to hit a fixed-wing aircraft with sufficient rounds to do damage is rather short, say 1,000 yards maximum. An RPG has to get this close to have any hope of a hit at all. Man-portable anti-aircraft missiles don't have much of a range either so the terrorist has to get up relatively close. Not only does the terrorist need to get close to his target but, similar to getting directly in front of a column of vehicles, it is much easier to score a hit on an aircraft if you are shooting from dead ahead or dead astern.

To secure against this threat, the flight path to and from a fixed-wing airfield must be guarded carefully by a cordon of infantry. This often changes according to the wind direction as aircraft generally take off and land into the wind when they can. This means you should be able to work out where the terrorist is going to be positioned, spot him and stop him. If an aircraft is coming in low to drop your supplies you owe the pilot the trouble of putting out a perimeter guard or whatever you can do to protect him while he is at low altitude.

The Parachute Regiment in the Falklands during the 1982 campaign to oust the Argentine invaders. Note the captured radar controlled 35mm Oilerkon anti-aircraft guns, bottom, and the prisoners, top, with (then) Lance Corporal Collington. (Photo Courtesy Robert "Bow" Collington, Retd. Sergeant, Parachute Regiment)

Aircraft depend on your protection when they are taking off and landing so don't let them down. Next time it might be you going home on R&R or tapping your fingers waiting for the air-strike to bail you out of a tricky situation.

## Choppers

Helicopters are much less vulnerable to small-arms fire than fixed wing aircraft, strange as this may seem. And I am talking about ordinary transport helicopters like the Blackhawk here as opposed to the flying tank called the Apache. A chopper is a relatively small target and can 'jig around' a lot more than a transport aircraft if the pilot thinks there is a risk.

Besides this, choppers do carry armour, often a great deal of armour, and they take a lot of knocking down.

I have flown in choppers when bullets have been cracking through the cabin and the skin was like a sieve – I've never had one let me down. All chopper crews are either heroes or crazy – I think it is in the job description. Many is the time chopper crews have come in under punishing fire to take out our wounded without a thought for their own safety.

The crew sit in armoured seats and the twin engines are surrounded by armour which is some help. The rotor blades are so tough they will cut the top off a telegraph pole. Seriously, I've seen this in the sales video. A chopper will run on one engine and, unlikely as you might

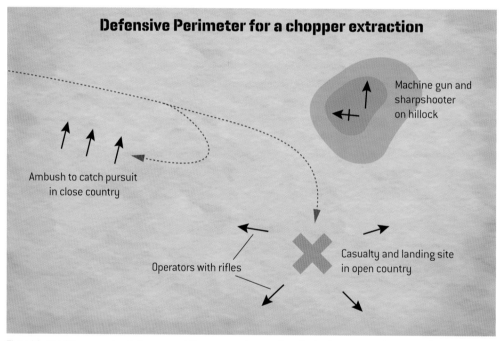

**Defensive Perimeter for a chopper extraction**

Machine gun and sharpshooter on hillock

Ambush to catch pursuit in close country

Operators with rifles

Casualty and landing site in open country

The object of the exercise is to prevent the enemy destroying the helicopter on the ground as it loads the troops. Where there is long distance visibility over open ground the riflemen just form a rough circle around the landing site facing out. The sharpshooter and the machine gun are placed on the hillock to prevent the enemy occupying it and give better range/visibility for them. The casualty is placed close to the landing site. Where the cover is close and visibility short range, as in forest or jungle, the enemy cannot see the landing site from a distance so the only way they can find their way to the position in time to attack it is by following the section. To avoid this the section forms a dog-leg ambush as they approach the landing site as shown at the top left of this diagram. In a hot extraction, where the section is in contact with the enemy, a gunship would accompany the extraction helicopter and strafe the enemy before a landing was attempted.

A group of prisoners are loaded aboard a UH-60 Blackhawk helicopter at Camp Korean Village, Iraq, 7 March 2006. (USMC)

A Merlin, with the back door down, and a gunner on watch, landing by a British Parachute Regiment patrol. (Photo courtesy Tom Blakey)

A Chinook loading troops en route to an operation in the Upper Sangin Valley, Afghanistan. (Rupert Frere © UK Crown Copyright, 2009, MOD)

think, if it loses both engines it can alter the pitch of the rotor in a manoeuvre called 'auto-rotation' and effectively glide into a bumpy, but survivable, landing from any height. I'm not going to tell you the sensitive spots for obvious reasons. The crew may get blast and small-arms wounds but when a chopper comes down it has either been hit by an anti-aircraft missile or been very, very unlucky. They are my favourite form of transport so I may be a little biased but choppers have saved my neck more times than I can count.

A chopper can, provided it is not overloaded, come in and go out steeper than a fixed-wing but taking off and landing are still the dangerous times. As you will not always be landing on an airfield from a chopper, when you do come down onto an area that is not secure get out there to the appropriate distance required by the ground and set up the defensive perimeter.

When you are waiting for a chopper to collect you make sure there is a good perimeter defence around where it is coming in to land. Don't use the same place regularly.

If you let one crew get shot up by idleness then the others may not be so quick or heroic in getting you out of a sticky spot. Protect the choppers with a perimeter defence and they will look after you.

# CHAPTER 6

# How to Avoid Having Artillery Spoil Your Day

For the purpose of this discussion I am going to define artillery as heavy guns firing shells of various kinds, mortars, rockets fired from the ground or from aircraft with various contents, and bombs dropped from aircraft with various contents. My reasoning is that at the receiving end of these it makes little odds how they get to you and the defence against them is much the same once you understand what they can do. It is also the case from a strategic planning point of view that ground-attack aircraft are just very long-range artillery with limited ammunition.

Operating these expensive, high tech weapons takes a great deal of brains and training. Defending against them does not. The principle defence against all these weapons – aside from a gas or chemical attack – is the simple hole in the ground. In conventional warfare any and all of these weapons may be deployed against you. In counter-insurgency warfare the main threats are rocket propelled grenades and mortars firing high explosive shells. I am covering the remainder because you never know what is going to kick off next and if you find yourself up against real artillery in a real war you will be better prepared.

# HIGH EXPLOSIVE BOMBS AND SHELLS

High Explosive is a chemical which burns very quickly and turns into a larger quantity of gas than the original volume of explosive. This means it expands suddenly and pushes the surrounding air out of the way creating a shockwave which pushes or mushes whatever it comes into contact with. In words this does not sound very threatening until you know

An Israeli Army 155mm mobile artillery piece fires into southern Lebanon from a position on the frontier, 13 July 2006. (Corbis)

about VOD. An explosive is defined as high, medium or low by its VOD or velocity of detonation.

The VOD is how fast the burn travels through the explosive and is expressed as metres per second. The idea is that the faster the burn the greater the shockwave. To give you an idea of what I am talking about here, a home-made explosive such as AMFO is a medium explosive with a VOD in the region of 2,500 metres per second whereas a high explosive such as PE4 has a VOD over 5,000 metres per second. If you had a strip of it 5km long and detonated one end then the burn would reach the other end a second later. Pretty quick hey?

A pure blast will kill you if it is close enough – and not-so-close I can tell you it feels like being kicked all over at the same time – but a small explosive charge with a little scrap iron around it

Shrapnel damage to a truck after roadside bomb attack in Baghdad, 17 April 2006. This was an IED rather than a shell but photo shows the shrapnel damage well and the principle is exactly the same. (Corbis)

can do far more damage to more people as the shrapnel travels in all directions like a thousand bullets. Shrapnel is the name given to steel or iron projectiles driven by an explosive charge. Sometimes the shrapnel may be ball bearings or nails but generally it is the splintered casing of the bomb or projectile similar to an old Mills Pineapple Grenade. There is also the prospect, with the advance of technology, for terrorists to be able to deliver all manner of nasty things by mortar or shell and I shall cover these in their place so you won't be living in happy ignorance.

The common shrapnel artillery shell or mortar bomb is a brittle iron casing filled with explosive and fused to explode in one of three situations to defeat different target types: on impact with the ground, in the air above the ground with a timer/radar or underground with a timer or delayed fuse. Timed/radar fuses to achieve air or underground bursts are tricky to use and not really a terrorist or even infantry weapon. They are usually employed by experts such as artillery units or the navy — although recently Western 120mm mortars have been issued with radar-fused air-bursting bombs. Nevertheless I will explain all three so you know what you are missing.

Shells from cannon and bombs from aircraft can also do all manner of clever thing such as splitting into smaller units called 'bomblets' as they approach the target but this is fine tuning to make sure everyone gets a piece and does not concern us here. The thing to remember is that when someone is firing explosive ordnance at you the best place to be is in a hole. Preferably with a roof.

## High explosive ground burst

All the mortar bombs that insurgents have access to, and most artillery shells, are set with what are called 'contact fuses'. All this means is that they explode when they hit something. This is usually the ground. Ground burst is good from your viewpoint — providing you can either get yourself flat on the ground a fair way from the point of impact or, even better, into a hole in the ground. A hole in the ground is the infantry soldier's second best friend — after his rifle — and his natural home.

If you cannot get on the ground when a high explosive shell lands near you then you are going to be absolutely riddled with holes from the shrapnel.

**A ground-bursting shell and a happy infantry soldier**

The good news is that you generally have plenty of time to react to incoming mortars and artillery so ground burst is not that much of a threat unless it's from a calibre of 155mm or higher. You won't have to hear mortars many times to recognize what calibre they are and how far they are coming from – so you don't have to rush needlessly.

Incoming artillery of 155mm and above sounds like a train coming into a station and is quite intimidating. The bang it makes is deafening but you are OK so long as you are in a hole and the shell is set to ground burst. Sometimes 155s will collapse your trench but fortunately they are fired from huge guns and insurgents cannot use them because they cannot be hidden from our air assets.

## High explosive underground burst

The underground burst shell or bomb works by having a detonator which is triggered when it hits the ground in such a way that it sets off a very short fuse. This fuse allows the shell to bury itself before it explodes and therefore has lots of packed earth around it to push against. The problem with an underground burst is that it is always a large shell and presumably aimed to collapse your holes and trenches. Your concern here, besides being blown to pieces, is premature burial.

Fortunately, in current conflicts at least, you are more likely to win the lottery than be subjected to an underground burst. This is because the shell has to be fired from heavy artillery which terrorists don't normally have – so don't lose sleep over it. If you did suspect the incoming shells were set this way you would be well advised to keep your head near the surface or lay in a hollow as then would take almost a direct hit to get you.

An underground-bursting shell and a buried infantry soldier

## High Explosive air burst

One of the biggest killers during World War I was air-bursting shells. Used against troops advancing in the open it is absolutely deadly. For obvious reason, one shell spreads its shrapnel out democratically so that everyone gets a share. It is even deadly if you are in an uncovered foxhole.

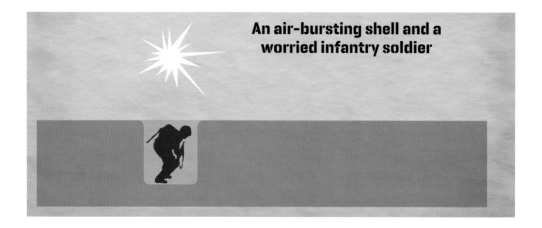

**An air-bursting shell and a worried infantry soldier**

Do remember that air burst is unlikely to be used against you on purpose. The one time when you need to be concerned is if you choose to site your foxholes under the shade of a tree to protect against the sun or rain. In this case any mortar round or shell which hits the branches of the tree becomes an automatic air burst. You will only do this once.

The way to deal with air burst, even air burst from a sheltering tree, is to put an earth roof on your holes. It takes some work but it is wonderfully comforting to have 18in of packed earth or rock over your head.

# NUCLEAR, CHEMICAL, BIOLOGICAL AND OTHER NASTIES

Most of the following toys are very unlikely ever to be used against you on the battlefield. Amongst the reasons are that they either take a great deal of trouble to make or they are very difficult to deliver properly – so they hurt the enemy rather than the user. Nevertheless I have covered the main exotic weapons groups to give you an idea of what is possible and what you just might face in counter-insurgency and other types of warfare.

## Nuclear weapons

A nuclear weapon has several ways of killing you. The first and most obvious is the blast which can flatten a city. There is also the flash which sets fires or fries everything within a certain radius. Finally there is the radioactive fallout which causes burns to appear on the skin like third degree sunburn followed by sickness, the breakdown of the body's systems and death. A hole in the ground the is the best answer to the blast and flash but a protective suit and gas mask is the only answer to the fallout. Following exposure there are drugs which can be administered to help the body survive and chemical washes which help remove contamination.

Nuclear weapons are used like conventional high explosive bombs in air, ground and underground bursts. The air burst burns and flattens a larger area but leaves little radiation to trouble occupying troops. A ground burst destroys hard targets on the ground over a wide area but contaminates that area. An underground burst is ideal for destroying the strongest of bunkers. Interestingly enough, the research in nuclear weapons is not to make them bigger and more powerful but to make them smaller and less powerful. Smaller so that they can be fired by battlefield artillery and less powerful so they can be used like super powerful high explosive just to take out a bunker complex, camp, or military grouping.

The good part is that nuclear weapons are incredibly difficult to make at all requiring huge amounts of radioactive materials and the machinery to concentrate this then the ability and science to do all manner of other things. Iran may have nuclear weapons by 2012, Pakistan already has them and may fall to the Islamic extremists shortly. North Africa is in a state of turmoil and the Islamic extremists, if they take over, may be able to obtain nuclear weapons from Iran. The days of superpowers threatening to flatten each others' countries are probably over but there is no guarantee a rogue nation will not try to launch a few missiles at a Western nation's cities perhaps from a submarine or surface vessel.

A terrorist group is not going to use a nuclear weapon in an insurgency campaign but they might be able to smuggle a nuclear device into a Western country. This would be extremely difficult to do and is more a concern for the security agencies than soldiers. The final reason that no insurgency campaign or country is likely to use nuclear weapons is that it would take the backing of a state or country to develop and deliver them. The CIA, MI6 and MOSSAD would immediately find out who they were because the material in the bomb has a sort of finger print which shows its origin and the US would nuke them back into the stone age. The usual suspects know that and though they may be fanatics they are not stupid.

Hiroshima, Japan, March 1946. This was a thriving city before a relatively small nuclear device was detonated above it as air burst in August 1945. (Cody)

## Dirty bombs

Dirty bombs are merely a high explosive charge with some radioactive material close by – perhaps around it. They are very easy to make as appropriate radioactive material can be sourced from medical facilities but difficult to hide and transport as detection is quite easy. When detonated the blast spreads the radioactive material as dust which, if strong enough, can cause radiation sickness. It is much more likely to do nothing or perhaps cause a handful of incidents of cancer in the long term. The main purpose of such a weapon is to close down a city for an extended period while it is decontaminated and to cause panic amongst civilians. Again, you are unlikely to find these on the battlefield in counter-insurgency warfare.

## Poison gas

Poison gas is quite difficult to make and even more difficult to place where it is wanted. It can be fired in shells or released from canisters but it is always difficult to be sure where the wind will blow it and how long it will remain sufficiently concentrated to be effective. The reason it is not used is that it is less effective than explosives or phosphorous in a conventional military or counter-insurgency situation.

## Napalm strike

Napalm is a mixture of petrol — gasoline for my colonial friends — and another substance about which I am not going to educate the amateur terrorist. The effect of the addition to the gasoline is to turn it into a sticky jelly which retains its burning ability yet does not evaporate quickly and can be spread around the target. If you come across this useful but unpleasant substance it will be a case of 'the biter bit' as it was popularized by US forces in Vietnam. Older folks will remember the pictures of the poor little girl running towards the camera with lots of skin burnt off by napalm. I'm glad she's ok now.

Napalm can be put into bottles and thrown but more usually it is put into aircraft auxiliary fuel tanks and dropped onto targets. Generally by our side. My only point in mentioning napalm is to tell you how to deal with it in the unlikely event you are at the receiving end. You may come across it in a riot situation or when your vehicle breaks down in a town and a crowd gathers. Water is no good and a blanket is not much better. A chemical fire extinguisher of the powder or foam type is about the best bet if you have one handy. Vehicle-mounted extinguishers are a comfort in times of need.

## Phosphorous

Phosphorous is a chemical which self-ignites and burns in air at a high temperature. You may have come across it as phosphorus grenades which are designated 'white smoke' allegedly for signalling. Our tanks have white smoke rounds also for signalling.... Yeah right.

UN sappers neutralize unexploded white phosphorous munitions in the southern Gaza Strip. Over 84 white phosphorous shells which did not explode had to be destroyed following Israel's three-week war in Gaza. The Israeli Army was doing a lot of signalling... (Corbis)

What it is really used for is clearing trenches and bunkers because when the grenade or shell goes off with a soft pop a cloud of white smoke appears and lots of little bits of phosphorous come flying through it. Whatever the bits touch they set fire to and if they land on your skin they will burn through until they drop out the other side. The white smoke, so good for signalling, also burns the lungs of anyone who breathes it so as a handy by-product it can be used to clear out bunkers more effectively than fragmentation grenades. When the blind choking creatures come out it is a kindness to shoot them. US artillery now have combination high-explosive and phosphorus shells which the gunners refer to as 'shake and bake'. Military humour hey?

## Anthrax

Biological weapons are things such as anthrax spores — bugs which start up infectious diseases. They are surprisingly difficult to make, to keep alive in transit and to deliver to the target. You may be aware that delivery by mail was tried with anthrax in the USA a few years ago and it did catch one or two people. If it can be done efficiently, disease spores such as anthrax make a terrifying weapons spreading panic across a city in hours. There is nothing to see and no one knows where they are or who has caught the disease. Nasty.

Anthrax comes in three flavours: you can get one type through a cut, another through eating infected meat and the third and most frightening, through breathing the spores. The symptoms start with fever and vomiting blood but it makes little odds as you are dead in 2 days as a rule. Sometimes it takes 60 days for the symptoms to show and sometimes less than 7 days. Imagine the effect of a news story where the people of a major city thought anthrax spores had been released. Hundreds of thousands of people would be potential victims and swamp the health services. The city would have to be sealed off and somehow decontaminated. A nightmare really. As a solider you are far more likely to act as security detail in a city with an anthrax scare than to come across it as an actual weapon.

# MORTARS AND HOW THEY ARE USED

It might seem counter to common sense but the biggest effect the terrorist artillery — and that is usually a mortar — is going to have on you, if you are following your SOPs, is to wake you up now and again. This is because in the sort of counter-insurgency warfare you are likely to be engaged in the enemy cannot use artillery or mortars efficiently — with aimed barrage fire – and all you need to defend against the individual mortar bombs they can throw at you is a hole close by.

Let me explain: properly set up artillery and mortars are arranged in batteries of three or four units and targeted in such a way that they rain down a terrific barrage of explosive projectiles onto a precise area for a couple of seconds – and kill everything in the area above ground. The correct response to this, apart from being in a deep hole, is to have some of your own radar-guided artillery set up to reply immediately onto the attacker's firing position. This is why so much modern artillery is designed to fire a few rounds then drive away before they get blatted. Fortunately, terrorists cannot use artillery or mortar fire against you properly because it would be vulnerable to our excellent air assets and counter-battery artillery.

Because of our side's potential for counter-battery fire, the best the terrorist can do is loose off the odd mortar bomb roughly in the direction of your hut and then try to get away before our radar tracks the bomb back to its source and replies with devastating results. So all you are likely to get coming at you is the odd mortar round. One or two at a time.

You will notice that all our main camps in unstable countries are spread out as much as possible so you have to be very unlucky to be hit by a random mortar bomb. In any event, the SOP is this: when in camp, make sure there are sufficient holes by all buildings and areas where soldiers gather so they can jump in when they are told or hear mortars fire. For your

An ammunition lorry, caught in the open, is hit by
enemy mortar fire in Normandy, 25 June 1944.
Mortar fire can be just as effective today. (Cody)

British soldiers are pictured firing flares from a 81mm mortar to illuminate enemy positions outside Basra, Iraq. (Allan House © UK Crown Copyright, 2006, MOD)

part, if you are a buckshee, don't argue or waste time, just get in that hole when you are told. That is all you need to do to stay alive.

The only time mortars can be a problem to you is if someone spots you on patrol out in the sticks and has a pop at you with their mortar. Remember they don't usually have the skills or equipment to produce a concentrated fire and a beaten zone – the kill area – so, again, you are going to have just one or two mortars shells coming down in your part of the world. You will hear the thud as the propellant fires the shell up into the air and then you will have some seconds before the shell comes back down again. Find a hollow if you can. Failing this just hit the ground near a big rock or wall and that halves the chances of you being hit straight away. Even in the open, so long as you are close to the ground, a mortar has to come very close to do you any damage. And while that is happening your very cross patrol commander is going to be calling in fire on the enemy mortar team. They have to be line of sight to hope to hit you and pretty easy to spot gathered around an 82mm mortar so if you have a drone up, or a chopper or even a sniper on a hill they are dead meat.

# CHAPTER 7

# How to Deal with Mines, Bombs & Booby Traps

If you remember nothing else from this book remember this: in counter-insurgency warfare, mines, roadside bombs/IEDs and booby traps kill more soldiers than everything else put together – 80% in some combat zones. The reason these weapons are so popular with terrorists is that they are cheap, easily available and one man can plant enough bombs, scattered around a wide area, to make life miserable for an otherwise overwhelming number of troops. The reason these weapons are so effective is that they are rather difficult to defeat. Difficult but not impossible if you remember your SOPs.

We have seen earlier how the main types of conventional, shop-bought, mines work. In this section we will look at how these mines are deployed in conventional warfare and how they are overcome. Then we will move on to look at how IEDs and booby traps work. Once we understand all these threats we can then look at how terrorists use mines, IEDs and booby traps in counter-terrorist warfare and, most importantly, how we can defeat them.

A friend of mine – the same fearless Taff Davies who saved my neck when I was hit by a grenade blast – was once part of a large patrol of 51 men. They were walking in single file, the regulation 5 yards apart. More than half the line walked over the mine without setting it off but then one was unlucky. We guess there was an AP mine placed on top of an anti-tank mine to make a big bang. The poor guy who trod on it, BEHIND Taff, went 50 feet into the air screaming all the way up and all the way down. The blast cut off both his legs at the hip, one arm at the shoulder and left one bone to the elbow on his other side. They only found the barrel of his rifle. Unfortunately, he remained fully conscious throughout the experience and lasted 90 minutes before he died. The flash from the blast had heat-sealed the wounds so he didn't continue to lose blood. Taff went to render assistance but what can you do? There was no chance of a casevac where they were and the best medics couldn't save someone in that state all those years ago. Taff took out a saline drip pack but while he was wondering where to put it, the victim used the bit of bone remaining from his arm to try to beat him away. It seems likely he wanted to be out of it as soon as possible which seems reasonable to me.

# CONVENTIONAL MINE WARFARE TACTICS

People usually think of the landmine in isolation as something a person steps on or a vehicle drives over with the obvious result. This is not what they were designed for originally and their use as hidden explosive devices is strictly outlawed by the Geneva Convention. According to

'Danger Mines' – children playing in Cambodia. (i-Stock)

the Convention, all mined areas must be clearly marked as such on the ground and on maps. There should be tape and signs to make civilians and troops aware of their presence and allow for later disposal. Why would anyone want to warn people?

The reason a minefield should be marked is obvious when you think about it. If a person knows there are mines in an area they will keep away from it. Simple really. And keeping people away by the threat of death is the ideal use of mines in most conventional situations. Let me explain. There are three ways in which mines can be usefully and legally used: to slow the advance of an enemy force, to funnel an attacking force into an area where they can be destroyed and finally to deny the use of ground to an enemy. This last may be problematic as we shall see.

In case you were wondering, I have avoided discussing mine laying patterns and density in case that were a help to our enemies. If you get to lay mines you will be told how and where.

## To slow an attacking force

You probably already know that the conventional way to assault the enemy is to get their heads down with suppressing fire and then rush at them to finish them off. If you are not

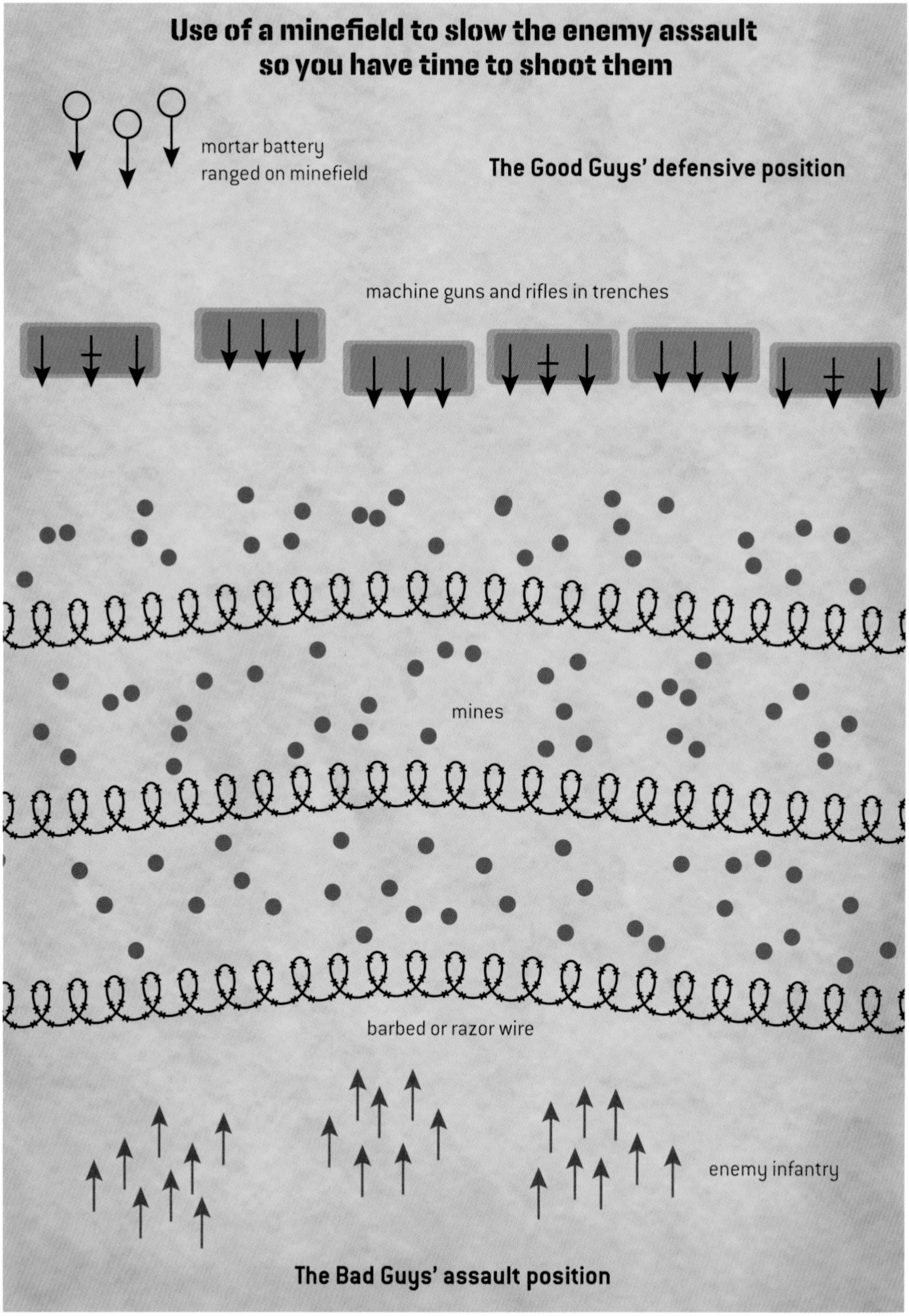

# Use of a minefield to slow the enemy assault
## so you have time to shoot them

mortar battery
ranged on minefield

**The Good Guys' defensive position**

machine guns and rifles in trenches

mines

barbed or razor wire

enemy infantry

**The Bad Guys' assault position**

fully aware of the principle tactics of infantry warfare don't worry as they are covered later. The point is that if you are on the receiving end of an attack the enemy is trying to keep your heads down with suppressing fire and come at you faster than you can shoot them down. They already outnumber you heavily or they would not be launching an assault so what can you to do slow them down while you shoot them?

There are two main ways of slowing down the advance of an enemy trying to assault your position large, or small. These are barbed wire and mines. Preferably used together for optimum effect. All you do is lay a pattern of mines and strings of barbed wire across the approaches to your position, make sure they are covered by fire, and then sit back and wait. When the enemy approaches, the mines and wire will certainly slow them down a great deal as they cannot approach faster than a leading man or vehicle can clear the mines. Slowing the assaulting troops gives you a massive advantage as they ought to be crossing open ground to get to you and you will be dug in like ticks on a hog with your guns laid on the minefield. Won't you?

## To funnel an attacking force

If you were attacking an enemy position you would spread out your men to make a less concentrated target. You should have been taught not to bunch up in tactics at squad level but the principle continues right up to Battle Group level. In small engagements when you bunch up you make a target for a machine gun. In a large engagement you make a target for artillery or air strike.

If you are on the defending side then you want to get the enemy to bunch up so you can kill them more easily. One way to do this is with a marked minefield laid out in such a way that by changing direction to avoid the mines, the attacking force is channelled into a killing zone which you have plotted for a strike by mortars, artillery or air cover. You will remember from the earlier section on avoiding artillery that all three of these weapons systems work best when the target is above ground. By definition, an attacking force is moving and above ground and if you can get them to bunch up where you want them then, for the artillery, it is like shooting fish in a barrel.

### Borders

It is totally acceptable, according to the Convention, to lay and mark a minefield on a border to stop civilians or military units from crossing it. Very often there will be guards to shoot anyone making the attempt and sometimes even to finish off the wounded. Landmines were planted

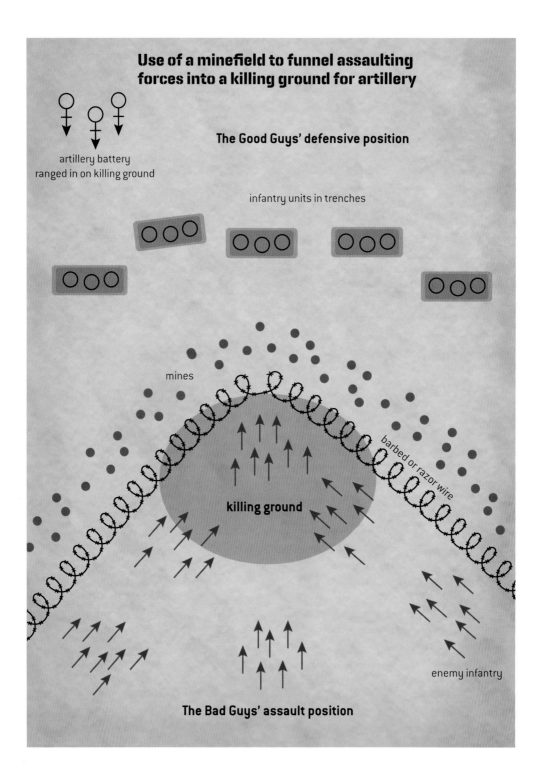

# Use of a minefield to funnel assaulting forces into a killing ground for artillery

artillery battery
ranged in on killing ground

The Good Guys' defensive position

infantry units in trenches

mines

killing ground

barbed or razor wire

enemy infantry

The Bad Guys' assault position

by the Communist East German State on the old demarcation line between Communist East and Capitalist West Germany. The mines were there to discourage civilians from venturing across from the poor East in search of a better life in the prosperous West. The guards used to leave the wounded and dead out in the minefields for a while as a discouragement to others considering a vacation to the bright lights.

## Protect fixed camps

You may have a major fixed camp – like Bastion in Afghanistan – where the concern is really about terrorists creeping up and taking pot shots at you rather than mounting a full-scale assault. In this case you can use a minefield to discourage the locals from coming too close under cover of darkness, fog or dust storm.

East German border with tower, fence, mines and the inevitable memorial to a failed escaper. (Photo courtesy Sergeant Roy Mobsby)

## Deny strategic points

There may be hills or other strategic points which you do not want to occupy but which you would sooner the enemy did not occupy either. An acceptable way to do this is to plant a minefield and mark it. In this case, you should take extra care to make the mines more difficult to detect perhaps including various booby traps and other entertainments. You might use this tactic when a hill overlooks your defensive position and you cannot station men there to deny it to the enemy.

## Deny enemy movement

In densely covered country, such as forest or jungle, the enemy may avoid tracks and use a different route every day to move men and supplies – thus making ambush difficult and expensive in terms of manpower. One solution is to spread landmines across the area using special artillery shells or by dropping them from aircraft. These particular mines are designed to be difficult to see in the dense foliage and to self-arm when they hit the ground. In dense cover they are difficult to avoid. This practice, of course, is highly unlawful besides being ethically questionable.

# CLEARING MINES IN CONVENTIONAL WARFARE

As you will now understand, there are two concerns relating to clearing a minefield in a conventional conflict. The first is that you have to find and neutralize the mines. The second is that you have to stay alive while the covering fire is trying to kill you. To get around these two issues there have been a load of techniques developed. Some are much cleverer than others.

## Avoiding the enemy fire

Given that clearing mines manually is slow and requires concentration it is probably as well that the enemy are not shooting at you while you are trying to do it. Even with most of the gadgets listed below people are exposed to fire. And of course, you don't want to lose your shiny new armoured bulldozer to an anti-tank shell do you? How do you avoid all this distraction? Put simply, there are two ways to stop people shooting at your mine-clearing team. You can either make them keep their heads down by bringing heavy fire to bear

Mine clearance: the loneliest job in the world. (USMC)

from artillery or small arms or you can lay down smoke to spoil their view. If you do use smoke, remember that defensive mortars, artillery and machine guns are likely to be set up on fixed fire patterns to cover the minefield without being able to see it.

## Defeating landmines

There are several ways of defeating a defensive screen of landmines and clearly some are better than others from the point of view of not getting you killed while you are doing it or later when you find the mines which were missed – the hard way. Here we will take a look at the pros and cons of the most popular methods.

### Electronic mine detectors

These gadgets are rather like metal detectors to look at and indeed they work in a similar fashion. Metal in the construction of a mine is picked up in the electromagnetic field generated by the machine and registers with a 'ping' or a twitch of the dial.

There are two things to think about here: these mine detectors only pick up mines which have metal in them so they will only find some mines as modern mines are not metallic and are designed not to be found. And think about it, if you had a box of old mines with metal bits in them wouldn't you scatter nails or something around?

There is research into mine detection equipment which finds mines even when they do not contain metal. I don't want to give too much away here just in case it helps our enemies but I understand scattering bits of plastic around is not enough to defeat these machines. The words thermal neutron backscatter imaging (TNBI) spring to mind as one such avenue of exploration. This technique measures the way the molecules in a substance vibrate and therefore can be used to identify certain specific materials. One argument against this is the idea of planting plastic chips as distraction but the TNBI can be tuned to pick up only a specific explosive or constituent thereof.

In the words of one 'techie' who is working on this: 'We describe an experimental and theoretical project centred on the use of nonlinear acoustic shock impulses and micro-electromechanical systems (MEMS) sensors to detect and image buried, regular-shaped inclusions at various depths in soil. The use of shock waves is partly motivated by the perceived limited capability of conventional acoustic delivery modes, and by recent research which shows that shock impulses travel efficiently in granular media.' Don't you just love having smart guys on our side? This geek-speak relates to a type of underground sonar but we don't have to worry about it until we get issued with the new kit.

### Pointed-stick mine detectors

Using a pointed stick or bayonet to probe in the dirt for a mine buried just under the surface will find every mine in a minefield because if the operator doesn't find it with the stick they will find it with their knees a short time later. This technique sharpens the concentration greatly and is surprisingly effective.

The stick technique is very slow as you may imagine. Every inch of ground has to be pored over to avoid missing one. It is no use under fire for obvious reasons but it is ideal for clearing areas sowed with mines by terrorists when you have the leisure time. Or someone has.

### Artillery fire

Concentrated high explosive shell fire will detonate a patch of mines. But it does take a thoroughly good 'stonking' to get most of them. I wouldn't want to go over the ground first.

### Bangalore Torpedoes

Captain McClintock, of the British Indian Army unit the Madras Sappers and Miners invented the Bangalore Torpedo at Bangalore, India, in 1912 and it first became widely used in World War I. It continued in use through World War II and up to the present day in Afghanistan where it is used for clearing booby-trapped areas. The latest model torpedo is manufactured by Mondial Defence Systems of Poole, UK, for the UK and US armed forces.

There are various designs but the idea is a sort of long stiff charge which is pushed across the target area from behind. When it is detonated it clears a path through. They are quite effective for clearing barbed wire, minefields and other booby traps. The range of effective operation is a strip about a metre wide and 15m long so a minefield may take several goes with some heroic Sapper following up each effort with a new torpedo. The shop-bought torpedo comes in sections with threaded ends each about 5ft long and these are screwed together from the safety of, say, a trench, and pushed ahead of the user towards the enemy.

British Royal Engineers and American Combat Engineers have also been known to construct versions of the Bangalore in the field by assembling segments of metal picket posts and filling the concave portion with Plastic Explosive (PE). The PE is then primed with detonating cord and a detonator and the pickets are then taped or wired together end-to-end to make a long torpedo, producing shrapnel that cuts wire and/or detonates a strip of mines. This method produces results similar to the standard-issue Bangalore, and can be assembled to the desired length by adding picket segments.

The newest evolution of the Bangalore is the Bangalore Blade, made from lightweight aluminium and using explosively formed penetrator technology to breach obstacles which the original Bangalore would have been unable to defeat. In a test detonation conducted on the television show *Future Weapons*, the Bangalore Blade blasted a gap roughly 5m wide in concertina wire, and created a trench deep enough to detonate most nearby AP mines. The Bangalore Blade

Improvized Bangalore Torpedo is placed amongst concertina wire during demolitions training prior to combat deployment in 2009. (USMC)

was developed in the UK by Alford Technologies and is intended for use with both standard army and SF units that require a lightweight, portable obstacle-clearing device. You will understand the benefit of a wider gap now won't you?

## Armoured mine ploughs

These are usually restricted to Armoured units or Engineers but a pair of large, angled 'Dozer' blades on the front of a tracked vehicle is perfectly effective against anti-personnel mines and pretty good against anti-tank mines too.

A mine plough. If the engineers can get one of these into action your life could be made a lot easier and safer. (Steve Dock © UK Crown Copyright, 2006, MOD)

## The foot as a mine probe

What I am going to say here may sound a little callous and thoughtless but it is true and simply a fact of life. It was calculated by the people who do these things – they are not infantry soldiers – that more men are lost to gunfire while picking their way through a minefield than are lost to the mines themselves. So the boffins decided it would be a good idea if the infantry charged straight through a minefield to get at the enemy. There are lies, damn lies and statistics. I know for a fact the Russian tacticians employed this tactic. All I can say is that they can go first.

I have found myself in a position where we infantry were considered far less valuable than the armoured cars we were escorting. We were obliged to walk in front of the armour along the dirt roads looking for the mines... This too sharpens the concentration wonderfully. My soldier friends in various parts of the world have used the locals to locate mines as you can be sure they know where they are. And in Aden another friend used to make the local head-man ride on the front of his ferret scout car and shout out when there was a mine ahead. As far as I'm concerned my job is to keep myself and my men alive.

# HOW IEDS AND BOOBY TRAPS WORK

The idea of a hidden bomb, in case it is not obvious to you, is that rather than dropping it from a plane or throwing it at you from a cannon or whatever, and risking being shot, the explosive can be placed in such a way that you will find your own way to it on foot or by vehicle and save the owner some trouble. Most hidden bombs in counter-insurgency operations are home-made affairs and referred to as IEDs – Improvised Explosive Devices.

When you have made your own way to the bomb it can be set off either by your standing on it or driving over it, in which case it is a landmine, someone watching from a distance can trigger it, in which case it is a command-detonated IED, or it can be set off by your own actions in which case it is still technically an IED but more usually referred to as a booby trap. Be aware, you do not have to be stupid or careless to trigger a booby trap – but if you are you will trigger one sooner. Depending on the theatre of operations and the technical skills of the locals you may come across more or less booby traps.

IED damage to a light armoured vehicle in Afghanistan. Happily the crew survived. (US Army)

## Terrorist use of anti-personnel mines

As you already know, anti-personnel mines for conventional warfare are made to remove your foot, make you unhappy and leave you a trouble to your friends. This makes sense to the military planners in the West but not to terrorists by and large. What they want to do is make the biggest possible bang and hopefully kill a bunch of soldiers as this makes news headlines. To achieve this they often place an anti-personnel mine, which can be triggered by your foot, on top of one or two much larger anti-tank mines which your weight would not normally trigger, or old artillery shells or other explosive items. These additives make one hell of a bang when triggered by the anti-personnel mine and will often take out other members of the team if they are too close. So watch your spacing.

## IED and booby trap charges

The explosive charge featuring in an IED or booby trap is not the most significant feature – oddly enough. The important part is how it is set off. The charge will usually be one or more mines, one or more artillery shells, or a big can full of home-made explosives as these are all far more easily available than shop-bought explosives. In any event, it makes little odds when they go off.

## IED command detonation systems

Some of the following I must necessarily keep a little vague so as to avoid training our enemies but I can tell you enough that you will be able to understand and, to some extent, avoid the threat. Specialist Explosive Ordnance Disposal staff excuse my simplification. I couldn't do your job.

The following are the most popular means of detonating an IED:

**Current down a wire:** Just like when you are using a Claymore Mine, a double wire is run from the charge to the place from where it will be detonated. This may be hundreds of metres. An electric detonator is attached to the wire at the bomb end and the detonator placed in the bomb. When the target approaches the bomb the ends of the wire are placed against the terminals of a battery and the detonator fires setting off the main charge.

**Radio signal:** If the insurgents can get the kit a radio signal is the easiest way to trigger an IED; but the right kit is a specific type of transmitter for the trigger end and receiver for the bomb end. As you may imagine, customs and transport companies watch out pretty carefully for where these are going and they are not on sale in Kabul High Street. Not openly anyway. A second best option is a rig similar to that used in radio-controlled models where the transmitter makes a little arm work called an actuator. A bit more work to set up but just as deadly. These toys are found all over the world.

**Mobile phone:** In areas where there is a good mobile phone signal a mobile phone is rigged to trigger a detonator when the phone is called and the detonator and phone placed in the charge. As the target approaches the bomb the bomber calls the number of the phone with the bomb.

## REMEMBER:

If you are searching or visiting a suspect building and there is a landline telephone you should never answer unexpected calls. Why? Because if I wanted to get you I would place a huge charge in the house rigged to be fired by a call to a mobile phone then keep calling the landline from a long way away until you answered. Then, knowing you were within range I would call the mobile...

**Detonation cord:** You will recall we spoke of Velocity Of Detonation (VOD) earlier? There is a material used in demolition which is called Detonation Cord or Det Cord. I am not about to teach terrorists how to use it here. Det Cord looks like a flexible white washing line being smooth and about 3/8 of an inch in diameter but unlike washing line it is made of an explosive with a VOD in excess of 5,000 metres per second. To all intents and purposes, when one end of the cord is fired the other end explodes pretty soon after. And it explodes with plenty of force to detonate a main charge as I have seen small trees cut down and limbs removed from prisoners with one turn of Det Cord. The technique for use is pretty simple: a detonator is attached to either end of the Det Cord and one detonator is placed in the charge. The other detonator is fired to set off the Det Cord as the target reaches the bomb. The detonation then travels down the Det Cord and detonates the bomb in a tiny fraction of a second.

## Booby trap detonation systems

The following tricks for triggering booby traps are by no means all there are, but sufficient, I think, to make you keep your eyes open and give you an idea of what to look for. As a general guide, pick nothing up, switch nothing on or off, don't open or close anything, don't step on anything and never answer a landline telephone call in a suspect building.

**Trip wire:** A wire string across a track or doorway can be easily rigged to pull the pin on a grenade or trigger a detonator.

**Door opening:** Opening or closing a door can be made to break or make an electrical contact thus firing a detonator.

**Light switch:** Switching an internal light on or off can be made to trigger a detonator in any one of a number of ways.

**Light sensitive:** Beware of opening doors on dark cupboards or shining a light inside as there are switches which are sensitive to light.

**Trembler:** There is a type of switch called a trembler which is principally a U-shaped tube filled with mercury or similar. When the item in which it is housed is moved the mercury slops around and makes a circuit which fires a detonator.

**Package lift:** It is not difficult to rig a switch underneath a package in such a way that it fires a charge when the package is lifted. There was once a spate of these rigged to boxes of cigarettes as a soldier could always be relied upon to pick up a packet of ciggies to see if there was one left inside.

# INSURGENT TACTICS FOR USING MINES, IEDS AND BOOBY TRAPS

You know now how mines work and how they are used on the battlefield in a conventional war. You also know how IEDs and booby traps work and that they are, by definition, only used in insurgency campaigns by the bad guys. To pull this all together we just need to look at how mines, booby traps and IEDs are actually used by insurgents to kill soldiers.

When you take over an area from the previous battalion you really must make it your business to discover what the local insurgents have been doing recently with regard to these three threats. If they have shop-bought landmines and are using them then you have to look out for these. If they are busy making and planting IEDs against road or foot patrols then be aware of that. If your work involves searching buildings or vehicles then make sure you have an idea of what booby traps have been used in the past – if any. Of course this is not perfect as the insurgents' supply situation may improve or get worse, and their tactics may change for some other reason, but it is a starting point and this awareness is something to begin tilting the odds in your favour.

## Insurgent strategy: Dead not wounded

Besides their lies and propaganda, the insurgents are trying to kill as many of you as possible to get your loving families to put pressure on your government to bring you home – and out of the insurgents' country. This is the whole purpose of their campaign and nothing else. It does not matter to them how many of their own side have to die to achieve this.

If you take a look at the news in any media you will see that though the death of a soldier nowadays doesn't make much a of a news story it still makes a much bigger story than a soldier wounded. Now contrary to what a lot of Westerners think, insurgents are not stupid. They may be crazy but they know that to make the news, and take another step towards getting rid of you, they just have to kill a few more soldiers this month than they did last month, however many of them you kill. So they are not out to blow your legs off – they want you dead. This is why they make big bombs and mines rather than using standard AP mines which just take off your feet.

As Ho Chi Min said during the Vietnam War, 'These Americans come over here, they kill many of us, we kill a few of them and then they go home.' And, though I loath Commies, I have to admit he was a wily old bird and absolutely right. The goal of an insurgency force is not to beat the invaders on the battlefield, it is to break the will of their politicians and civilians back home. The Americans managed to kill around 3 million Vietnamese without breaking their will but the American people forced the withdrawal of American forces when the US death toll reached 58,000. You can do your part to achieve victory over the insurgent by staying alive and on your feet.

## Landmines

The main difference between mines and IEDs from a tactical point of view is that mines do not need an insurgent watching them to kill you whereas a command-detonated IED does. From an insurgent's point of view this is either good or bad according to the situation.

A mine does not need watching, or a command wire to give it away, but it has no discrimination and will kill the first person to walk over it be that civilian or soldier. This means a mine can only be placed where a soldier will walk, rather than a civilian, or the locals must be told. If the locals are not told then this will waste the insurgents' mines and possibly upset the locals, so locals are always an important source of information for you. In addition, mines are patient and will sit there, effectively forever, waiting for their target without risk to the insurgent. An IED, on the other hand, can be placed anywhere, regardless of passing traffic, and the person detailed to detonate it can allow any number of civilians to walk or drive over or past it safely until the chosen target arrives. Of course this ties down at least one man, and puts them at risk of being shot, but it does allow the weapon to be selective and detonate without being wasted on a civilian. This is why IEDs are so popular with insurgents.

### Avoiding AP mines

Insurgents want to plant their landmines where they think you will walk of course. But mines can only be planted in soft ground. Insurgents don't generally have lots of time and they don't generally have an endless supply of mines either so they try hard to make each one count. And they don't just want to take your foot off as we saw above, so there is a good chance they will plant extra explosive with an AP mine or perhaps an anti-tank mine underneath. Not just to make sure of the guy who treads on it but to get his mates too.

There are no sure ways of avoiding mines, bar staying in camp, so you do all you can and hope for the best. Most often insurgents don't want to kill the local villagers, goat herds or whatever, as these support them with food and information regarding your movements, so they either don't plant mines near the villages or they tell the locals where they have planted

## TOP TIP!

### How to avoid mines

Mines must have a soft surface above them which can transmit your weight and press down on the mine to cause detonation. The mine also has to be buried. This means they will only be laid in soft ground, as opposed to asphalt, so stick to the hard stuff when you can. If you see a patch of broken asphalt, walk around it.

Mines are often left in place for an extended period so the earth may blow away from the top and make the prongs or whatever visible so keep your eyes open. If the mine has been planted recently there may be signs of disturbed earth so step around that too.

It is always a good idea to avoid bottlenecks on a track where a mine-layer will know you have to step. Now I have mentioned it, you will notice how on bends, by walls, by trees and at the approach to bridges tracks tend to funnel people over the same narrow strip of ground. Don't walk on it until you see a local walk there first.

mines. Around a village, by far the best way to avoid mines is to persuade a few locals to walk in front of you. Of course your commanders may not allow this so instead watch the local's movements carefully before committing to a route.

If mines are expected, and you can't use the locals to find them, then it may be worthwhile having a mine detection device at the front of the team but if this is not practical then the front man should keep his eyes open for signs of a mine, rather than enemy attack, and the next man protect him but keep well back. Change the lead man regularly unless you owe him money.

When it is your turn to be in front of your patrol, or 'point' as some say, you need to keep your eyes open for a couple of things: freshly disturbed ground which might suggest a recent planting or the prongs of the mine sticking above the surface where wind or rain or bad laying has led to the business end being uncovered. Good luck.

## Anti-tank or vehicle mines

The principles are very similar to anti-personnel mines in terms of tactics but these mines often have even more explosive added to defeat our increasingly heavily armoured vehicles.

### Avoiding anti-vehicle mines

When you are on the road you are relatively safe from mines (obviously not from IEDs) but when you go off road you are vulnerable. The good news is that with a vehicle you have some protection. Of course if you blow up enough explosive under any vehicle it will kill everyone inside but the mine protection on vehicles is getting better and the chances are that you will get away with it most of the time.

The down-side is that you cannot look at the ground when you are travelling at vehicle speed so mines are hard to spot. Stay on the road when you can. When you have to go off road try not to follow a track and head cross-country. Avoid bottlenecks at bridges, fords etc as this is where the mines will be. By far the safest way is to get locals to hitch a ride – and failing that keep an eye on where other road users go and drive in their tracks.

### Action on detonating a vehicle mine

If your vehicle hits a mine, and remember you cannot know at the time if it is a mine or an IED, then take action as if you are going to come under fire as this often follows an IED strike. Someone needs to call in a contact report so your boss isn't worrying about you and you may even get some support or a casevac. The main injuries from explosives going off underneath armoured vehicles is to your feet and lower legs from the shockwave hitting the floor and your spine from the vehicle jerking rapidly upwards. In view of this there is an argument for keeping your feet off the floor and sitting well strapped into a sprung seat to save your spine.

Depending on the vehicle damage, casualties and if you have other vehicles you may choose to jump out and take up a defensive position, return fire from an armoured vehicle which is unlikely to be hit by another mine as it is not going anywhere, or you may jump into your other vehicles and make off. If you have to run away make sure you take or destroy all papers, maps and personal possessions. Then destroy the vehicle you are leaving as the opposition will certainly search it and take anything useful.

## IEDs

As you probably already know, an IED can be planted almost anywhere – it can be in soft ground or placed in a tunnel under a road. There is little difference between an IED aimed at foot soldiers and one aimed at a vehicle except possibly size of charge and that if the enemy are clever enough they may be able to make a shaped charge to give better penetration of the armoured vehicles underside.

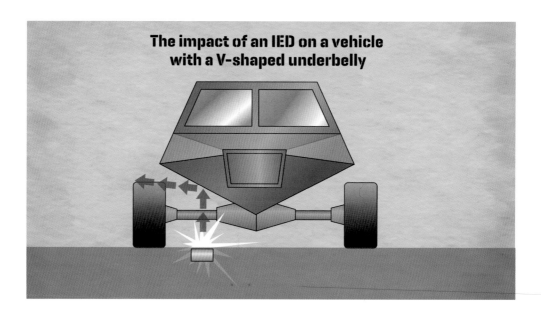

**The impact of an IED on a vehicle with a V-shaped underbelly**

## Avoiding IEDs

This is not easy but it is one of the most important skills the soldier will learn in this book. The first few tips are for the ordinary soldier to improve the chances of his team. The real counter to IEDs is high tech and mentioned below as far as is safe to tell:

### Soldiers

- Keep your spacing
- Look carefully for detonation cord and detonation wire
- Watch the locals and if they clear the area go with them
- Try to stay off the regular path when you can – as with mines
- In open country send a party ahead on either side of the main patrol with the aim of flushing out the operator of the IED or finding the Detonation Cord or wire
- Drive quickly: be aware that moving at speed makes it more difficult to trigger the IED under your vehicle

### Technical

- There should be equipment at the front of a patrol to block telephone signals in the area and therefore prevent a mobile phone from sending a signal to detonate
- There should be a radio transmitter of a type which transmits pulses over a certain range of frequencies to advance detonate any IEDs intended to be triggered by a radio signal
- There should be a type of transmitter which produces electromagnetic pulses to induce a current in local wiring so as to advance detonate any IEDs intended to be triggered by a wire and current

### Vehicle protection

All things being equal, the more armour under a vehicle the bigger bang it can survive but this quickly puts you in the situation where the vehicle cannot move at any speed. I have seen a main battle tank turned over on its top like a tortoise by a large IED.

You might think a bomb under a motor vehicle is just a bomb but not so. Mines always explode under the wheel or track of a vehicle and to a great extent this channels the blast away from the crew cabin for the loss of a wheel. Which is a pretty good bargain in my book.

An IED, on the other hand, can explode anywhere under or near the vehicle. This means that an IED with a shaped charge directing the blast straight upwards can be very effective against the crew of a vehicle.

To cut down on weight while maintaining protection it is a good idea to use vehicles with a V-shaped under-belly. I used vehicles of this type manufactured by the South Africans more than 30 years ago and the armour sloping away from the blast – be it mine or IED – really does cut down on both penetration of the vehicle and the lifting of the vehicle which causes so many injuries to feet, legs and spine.

### Action on detonating an IED

When there is an explosion amongst your patrol or your convoy you cannot know immediately if it was a mine or an IED. A mine is a one-off event and there may not be an insurgent for miles or any further problems but an IED certainly has at least one person watching, as they triggered it, and there may be a whole ambush party. Worse, there may be another IED placed close to the original blast just where your people will rush to take up a defensive position or where you are likely to set up a command position while you give first aid.

Your first problem is what to do about the casualty – and this is not an easy decision. So many men have been killed by the second blast while going to the aid of their wounded comrades that you really should leave him until the area is secure. If he looks dead, certainly do not approach his body. Many men feel obliged to go to the aid of their wounded mates, I can understand that and they are heroes for doing it, but it is a bad tactical decision so make sure no more than one man goes to the casualty. And be aware you may lose him too.

The rest of the patrol should take up a defensive position as if coming under fire. But be aware that this is exactly what the insurgents expect and so you should avoid obvious positions like ditches as these places may be mined or have further IEDs waiting.

## Do not think this will not happen to you

I am going to tell you a story now so that you remember. Every word is true – my mate 'Bow' Collington of 2 Para was there and you can check the story out on the internet or in records. In August 1979 a number of British Paratroopers were driving in a convoy of 4-ton soft skinned trucks near a place in Northern Ireland called Warrenpoint. As they passed a road-trailer the IED hidden within it – 400lb of Amfo fertilizer explosive – was detonated by radio signal, using

Sniffer dog Panchio (Arms Explosive Search) and handler take a break out of the sun. (Photo courtesy Corporal Natasha Mooney, 104 Military Working Dog Sqadron)

a model aeroplane transmitter, destroying the rear vehicle and killing six Paratroopers. Without electronic equipment there was no way to avoid this bomb.

The Paratroopers debussed and took up a defensive position firing back at snipers over the Irish border. There is some argument over whether they were firing back at snipers sited over the Irish border or reacting to ammunition cooking off after the blast. The fact is that the Irish Police did find forensic evidence of firearm use on the hands of two IRA members arrested shortly afterwards and suspected of being involved in the blast. On hearing the explosion a nearby patrol of Royal Marines radioed the contact in and further Paratroopers were sent to the scene of the blast to provide reinforcements. A colonel was despatched together with medical staff by helicopter to take command of the incident and help the injured. Arriving on the scene he set up his command post in a nearby isolated building with a wall around it— a gate lodge — accompanied by other officers, medics and guards.

For some time the IRA had been watching British Army's tactics in reaction to IEDs and expected the officer commanding the situation to set up his post in the gate lodge. **This isolated building was the very reason for the siting of this attack.** Just 32 minutes after the first IED was detonated a second and larger IED planted at the gate lodge many days earlier was detonated killing a further 12 soldiers including the colonel and some other officers. The moral of this story? Don't use obvious isolated cover after an IED attack. Indeed avoid obvious positions at all times as a sniping attack could be used to drive you onto an IED.

## Booby traps

Because these rely on your efforts to set them off they are generally set up in buildings or vehicles. Any type of building or perhaps a parked car. The reason is pretty obvious; patrolling outside you are going to touch nothing but the ground, so a mine is used, whereas if you have to look in a vehicle or building you might be tempted to touch something and trigger a booby trap.

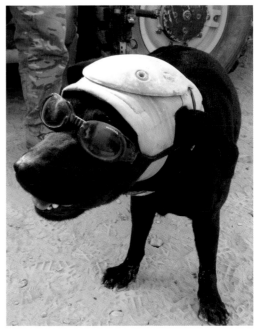

Panchio ready for the party with scratch hat and army issue 'Doggles'. (Photo courtesy Corporal Natasha Mooney, 104 Military Working Dog Squadron)

### Avoiding booby traps

To go into a building you have to open the door, tread on the piece of floor inside, then perhaps either put on a light or shine a torch. All of these can be set to trigger an explosion. Because the triggers are often simple and manual, as opposed to electronic, you cannot use a clever electronic gizmo to set off booby traps. By far the best thing would be to send a local in first and have them do all the things which would set off a bomb, but this would not be allowed by the rules of engagement in most Western armies. Failing this, blow open the door and use sniffing equipment or an unfortunate dog to check for explosives. If you have to go in, go in through the window to avoid pressure sensors. Fortunately, just like mines, booby traps are not very selective so they will only be placed where the locals do not go — empty or derelict buildings — so use your head and think if the opposition would have been sure the place was unused. I am not saying the insurgents will lose any sleep about killing a local but they generally won't want to waste the explosive or electronics.

# CHAPTER 8

# How to Deal with Suicide Bombers

When you operate in an area affected by an extremist insurgency you will very likely come across suicide bombers so keen to kill you that they are prepared to destroy themselves. In this section I am going to explain how they rig up suicide bomb kits, how they use these bombs and how you can defend against them.

We saw earlier how a missile, artillery shell or bomb dropped from an aircraft is, in principle, just a way that an explosive weapon — a bomb — can be delivered to a given target and that the insurgency variants of these are fairly crude and ineffective owing to lack of technical assets.

We have also seen how a landmine, IED or booby trap is an explosive weapon — again a bomb — which can be laid, hidden, where the enemy, you, will come of your own accord. These are somewhat more effective.

The suicide bomber is a third class of weapon delivery system. It is effectively another explosive weapon with a delivery system which gives it the most dangerous attributes of both the two weapons above: the suicide bomb can be delivered, invisibly and silently, to a precise chosen target and detonated at a time of the attacker's choosing. Like an invisible guided missile.

Besides generating the reasonable fear of a sudden attack, the suicide bomb has one other attribute which works in its favour; its effect on the Western mind. We Westerners find it difficult to understand the mindset of a person who will choose to kill themselves in order to kill some of us. This terrific dedication or hatred makes us uncomfortable and weakens some people's resolve to fight them.

**As a soldier it is simply important to understand** that people can be made to believe the most crazy things and, through holding these beliefs, be persuaded to kill themselves and others. Understanding this takes the psychological sting out of their actions.

The fact is that all religions have a series of basic techniques built into them which press the buttons we all have in our minds and build belief in the individual member. These techniques include meeting with other believers for reinforcement, singing songs which make statements of belief and performing acts which would only be performed voluntarily by someone who already believes — such as giving part of your wages to the priest. These actions, and others, build belief in any human mind by pressing its buttons. A person exposed to these techniques over time has no more choice in what they come to believe than a computer does when you programme it. And of course it works far quicker and stronger on children.

The lack of education, the poverty and the well established religious doctrine that exists in all areas where insurgencies are currently occurring makes for a potent mix. Religious leaders of all kinds have always promised that obedience to their commands will improve what happens when you die. Given you have total belief in that promise, dying as a martyr, and leaving a hard, poverty stricken life, in return for a certain ticket to paradise and lots of pretty girls, is quite good deal. On top of that, if you are a boy growing up in poverty then the promise that the mosque will give money to your family when you have died a martyr and gone to heaven is not going to put you off is it? So suicide bombs are an understandable reality for anyone on current operations, the important thing is to be prepared.

# WALKING BOMBS, VEHICLE BOMBS AND INSURGENCY TACTICS

The suicide bomber is a different order of threat to other forms of attacker: ordinarily a person who is going to have a go at you wants to either win a shooting match and survive or at least get away after doing some damage and, again, survive. If they are a reasonably sane bomber then they generally want to plant the bomb with a timer or trip and get away.

To make sure we understand one another, consider this: a bomber walks into a restaurant and places a rucksack bomb down by the coat rack with the timer ticking and set to 10 minutes so he can escape. He turns and begins to walk out. Someone shouts and he runs. He gets away. Certainly the rucksack is now a threat. Some hero – and there is always one – grabs the rucksack and runs outside with it, dropping it down a drain to muffle the blast. A hundred lives are saved and a bomb wasted.

The same bomber walks into the same restaurant with a rucksack bomb on his back, he detonates the bomb. A hundred people die. If you look at this from a purely tactical point of view, it means anyone who wants to survive an encounter must always limit their range of activities to those from which they might reasonably expect to escape. They will only consider a plan that gives them a reasonable chance of survival and more often than not this means they will attack you less effectively.

Anyone who is prepared to die to kill their enemy, on the other hand, is a much more serious threat. If a person is intent upon dying on the mission then they cannot be stopped by threats of retribution or risk of capture. They are also not limited to operations where they have any chance of escape at all. Anywhere to which they can gain access is an acceptable and attainable target.

## General tactics for suicide bombers

First of all you must realize that the suicide bomb is in principle a terrorist weapon for use against civilians. I say this because in a military conflict you would shoot a suicide bomber running towards you like you would any other assailant. Where a suicide bomber comes into his or her own is for attacking groups of civilians shopping, dining, watching a play or working in government offices. By this type of attack the terrorist/insurgent group spreads terror amongst the civilian population, disrupts the running of the country and brings about calls from the public to give in to the insurgents' demands. We are looking at suicide bombers because, you, the soldier, will be called upon to protect civilian targets, both at home and abroad, and therefore stopping the suicide bombers before they reach their target. When you become Special Forces then the tasks just get trickier. This is not a good job, in the general scheme of things, but we will see how best to achieve it and stay alive shortly.

The use of a suicide bomber as a delivery technique has two advantages over a planted bomb from the point of view of the terrorist/insurgent cell leader: firstly it is very likely that the bomb will not be detected by the casual observer and disarmed by the security services before it is detonated. Secondly, the non-suicide bomber has the hope of escape, and he/she may choose not to place the bomb in the ideal place on the spur of the moment, substituting a safer location, and that choice can spoil an operation from his/her handler's point of view. In every way therefore it is better that the bomber has no hope of survival and they will then not spoil or limit the operation by being tempted to try to allow for their own escape.

The tactics employed by the handlers of the suicide bombers vary slightly according to the situation. Because non-suicide, hidden bombs can be placed pretty much anywhere lacking high security the field is pretty much open to what the insurgents want to achieve. But after the first hidden, timed, bomb attack the area of operation will become security conscious with measures taken to combat both planted bombs and suicide bombers. This reaction, forced upon the local government, immediately gives the terrorists/insurgents an early victory by costing their enemy – the state – money and political strength. But these initial

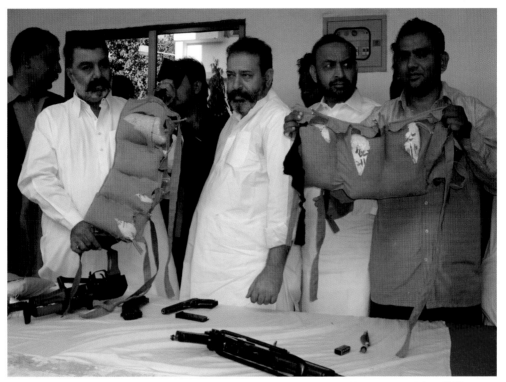

Pakistani security officials show suicide bomb vests, arms and ammunition recovered from Taliban militants during an operation, in the southern port city of Karachi Pakistan on 30 November 2010. (Corbis)

reactions, such as extra vigilance, removing waste paper baskets from public places and checks on the entrance to public buildings, will make the future planting of timed bombs in the most effective places much more difficult and bring forward the use of suicide bombers which are harder to detect and to stop.

## Walking bomb

The problem with a walking bomb, from our point of view, is that it doesn't take a lot of explosive to do a lot of damage. It is not difficult to hide 20lb of explosive covered in ball bearings fitted into a sort of vest and hidden under a coat without looking too suspicious. Compare the damage potential of that package with the 1oz of explosive in a hand grenade. And you can get a lot more explosive than that in a rucksack.

## TOP TIP!

### Don't become desensitized

If you have ever done checkpoint duty will know already that people come in all sorts of shapes and sizes – and most of them are doing nothing wrong at all. I would even go so far as to say that the most suspicious characters are probably innocent and the most innocent looking are likely to have a bomb. You wouldn't send in someone looking suspicious with a bomb would you?

No, you would send half a dozen suspicious characters through the checkpoint spread out through the day. The guards would see their shifty eyes, nervous movement etc and check them out carefully. By the end of the day, end of the week or end of the tour the guards would be sick of looking too closely at anyone who looked wrong. The target has been desensitized. Then the real bomber would come in looking like sweetness and light – just before blowing everyone and everything to Hell. Moral of the story: never relax your guard and remember that appearances are often deceptive.

### The vest bomb

This is the traditional hidden bomb beloved of terrorists in the Middle East and elsewhere. The charge is generally limited to around 20lb of explosive and this charge is often covered in ball bearings to increase the shrapnel produced at detonation.

### The rucksack bomb

The rucksack bomb is favoured in low intensity areas such as Western cities where a rucksack is not immediately assumed to be a bomb. Several of these bombs were used in the attack on the London subway on 7 July 2005. The advantage is that a rucksack can contain much more explosive than a waistcoat. The only real limit to the use of these bombs is the ability of the terrorist to obtain or manufacture explosive.

## Suicide bomb design features

All of the following concepts apply to the detonation of walking bombs. Some of them apply to vehicle bombs too:

**Suicide trigger:** All suicide bombs are set up so the carrier can set off the bomb when they feel the moment is right. If you stop a bomber and it looks like they will be prevented from reaching their actual target they will certainly trigger the bomb to take at least the search crew with them.

**Dead man's handle:** Very often, bombers are wired with a 'Dead-man's handle' so that if they lose consciousness through being shot the bomb still goes off.

**Remote detonation:** Just to make things a little more interesting, the bombs carried by suicide bombers are sometimes rigged for remote detonation by a watching handler just in case the carrier gets cold feet. The mechanism is similar to that for a radio or telephone detonation of an IED so the handler needs to keep his eyes on the carrier – presumably from a safe distance.

**Poison:** There is a suggestion that some bombers may be given a slow acting poison, along with other drugs to increase their compliance, before a mission. Perhaps on the pretext that they may be captured. In reality it means they have no hope of survival and might as well use the bomb efficiently.

## Vehicle bomb design

There is little to say about vehicle bombs other than that their size can be enormous. They may be carrying a charge varying from several hundred pounds to several tons of explosives. No special vehicle is required except that it be sturdy enough to carry the weight of the bomb with one exception: if the target is guarded by barricades a vehicle may be selected which has the ability to crash through the obstacles for instance a heavy truck or even mobile plant.

To make an effective attack with a vehicle bomb, due to the sheer power of the blast, it is often enough to park it in the street outside a target building. I am not going to teach terrorists about the effect of shock waves from bomb blasts so ask your friendly demolitions sergeant.

# DEFENDING AGAINST THE SUICIDE BOMBER

As a soldier, no matter if you are infantry of the line, SAS, Ranger or SEAL, you are very likely to find yourself guarding some person or place from terrorist attack by a suicide bomb. As a Special Forces soldier the assignments are just to trickier places and more valuable targets.

Someone has to do it and you had better know how to do it right as the people who die when it is done wrong are generally the guys doing the guarding, i.e. you.

The reason I say this is that the only sure way to keep a bomber out of a building or an area is to set up gates on all the entry points and search the people or vehicles coming in. There are many ways of doing this and it is not rocket science to keep the people inside totally safe from a suicide bomb attack if you check everyone trying to get in.

What is slightly more difficult is keeping the poor sods doing the searching alive because a suicide bomber who is stopped for a search and prevented from reaching his goal knows he will not be able to escape and is very likely to take out the people doing the stopping if he possibly can. Better to kill a few unbelieving soldiers than no one at all. Staying alive while stopping bombers entering a secure area does take proper planning and organization, and may require you to be a little ruthless to stay alive yourself. That is what we are going to look at now.

Depending on where you are working there may be different rules of engagement. What this amounts to is that the politicians may prefer that you take more risks and stop you

Two suicide bombers attacked a bus carrying Afghan Army soldiers on the outskirts of Kabul, Afghanistan, killing six soldiers on 19 December 2010. Suicide bombers will continue to play a role in the war in Afghanistan so be vigilant. (Corbis)

shooting potential bombers to make the politicians look good; or you may have the freedom to shoot anyone who comes too close and doesn't look right. Depends on your politicians.

For propaganda purposes the bad guys may also send non-bombers – women and children – to act suspicious, get too close and perhaps get themselves shot to make you feel bad and lay off before sending in the real bomber. Wise up to this tactic and don't let it happen.

The two main types of suicide bomber you might have to contend with, the walking bomb where a pedestrian carries a load of explosive on his body and the motorized bomb where the bomber drives his vehicle packed with explosive into the target area before detonation, both demand very similar defence techniques. The difference with vehicle bombs, of course, is that they tend to be much bigger, so they don't have to get quite so close to do damage, and they usually have to stay on the road.

I will cover walking bombs first as most of the concepts apply also to vehicle bombs, then I will cover a few principles which apply just to vehicles. Like almost everything else in this book, it is mix and match. Get the principles inside your head, keep your wits about you and use what serves.

## Walking bomb defence

**Control entry:** The first, last and only step which will prevent an effective suicide bombing mission is to control access to the area which is at risk from attack. Put a ring of steel around it. You will understand that there are huge numbers of such targets in every city and they cannot all be defended. Areas to be secured could be single buildings with access control points or the whole of a town centre with a fence around it. But that is not your worry; there are people way over your pay grade whose job it is to decide which targets are so important that they set you to defending them.

To stop any and every bomber you just throw a cordon of some kind around the area with guards on the entry points and search every single person trying to enter.

### REMEMBER:

When the insurgent knows you are searching every single person going into a certain area then they will not send in a bomber as the mission is bound to fail in its primary intent and only kill a guard or two. If you only search every third person the bomber's handler may take the chance. So make sure you search everyone OK?

Often these points are access doors to a building but they can just as easily be gates in a wired off market area. Once the cordon is up the people inside are safe from suicide bombers. So long as you make sure no one tunnels in.

A monkey can find a bomb on anyone passing through a checkpoint and prevent it killing the people inside the target. Unfortunately, that monkey is going to be the target of last resort for the bomber and, worst of all, that monkey is going to be you. The remainder of this chapter, therefore, is not so much about how you stop suicide bombers, you know that already, it is about **how you stop suicide bombers and stay alive yourself**.

**Keep everyone away from your checkpoint:** So what can you do to stay alive while preventing a suicide bomber gaining access to the area you are securing? The guiding principle behind dealing with all suicide bombers is to keep the people and vehicles too far away to kill you until you know they are not trying. How do you do that?

The ideal is to have people form a queue some distance away from your Search Point. This can be achieved quite easily using razor wire to force people into a line – similar to the way banks do in the West – though most banks use shiny little posts and cheerfully coloured ropes rather than razor wire. At the head of the queue should be a line on the floor, called a 'kill line', and a sign in a suitable language telling visitors that if they pass the line without being called they will be shot. Close to this position you should site an electronic sniffer to pick up explosives and a body scanner and video camera if you can get them. It is very difficult to make or carry explosives without getting some trace of the scent on you and this will catch almost all bombers and a few bad guys who have been handling explosive too. The civilians don't have to know about the body scanner which will show anything they have stuffed inside themselves. Anyone with bulky clothes can be asked to open up for the camera.

The gun and search teams will be sufficiently protected to withstand any explosion occurring behind the kill line. From behind the safety of the kill line people are told by the gun or search team to advance one at a time to a Search Position shielded from the queue by blast walls. If anyone advancing looks suspicious or bulky then have them open their clothes before approaching the search area. Because anyone trying to rush the Search Position from the kill line will be shot by the gun team the only weapons reaching this point should be possibly small knives – and the Search Team do have weapons and full body armour. When a person is cleared here they are allowed to advance past the kill line and enter the secure area.

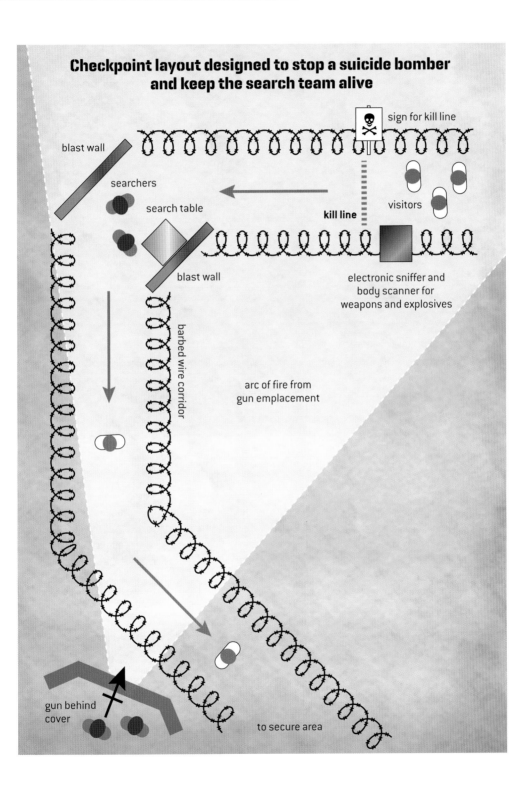

# Checkpoint layout designed to stop a suicide bomber and keep the search team alive

sign for kill line

blast wall

searchers

search table

kill line

visitors

blast wall

electronic sniffer and body scanner for weapons and explosives

barbed wire corridor

arc of fire from gun emplacement

gun behind cover

to secure area

About the best thing you can do when you have to do physical searches is make the visitor stand with their arms away from their body so they cannot trigger a bomb so easily. Then have the searcher approach the visitor from the side while a marksman aims at the visitor's head with orders to shoot if they make any sudden movements. This might just prevent a bomber triggering the detonation.

It may be the case that you are prevented by someone who 'flies a desk' from dealing with the situation as safely I am showing here on account of upsetting the locals or his boss' voters. You may even have to send someone forward to do the search. In this case draw straws for 'point' and change over at intervals to give everyone a fair crack at finding a bomber the hard way.

It is not a bad idea to set up your coral in such a way that an approaching bomber is not trapped in a queue before he can see how carefully you check everyone. This way he may just go away and trouble someone else. Better still, you might give anyone leaving the queue a careful once-over.

## Vehicle bomb defence

The guiding principle behind dealing with vehicle bombs is also to **keep them at a distance**. The problem is that a truck can carry so much explosive that it doesn't have to get very close to you or the target to do a lot of damage. It also takes some stopping. If you stop a vehicle bomb personally at a check point you are almost certainly going to die and there is nothing you can do other than go to mass regularly or carry a lucky rabbit's foot. In the accompanying diagram, the distances between each station should be as great as the area allows.

Fortunately, the chances of this happening are fairly remote as there is a

## REMEMBER:

Suicide bombers can be prevented from detonating their bombs in any chosen area given the political will to restrict freedoms and the military resources to do the searching. Normally the areas where these tactics can be applied must be restricted to relatively high value targets such as aircraft facilities, government buildings and military posts. Do your job well and you'll stay alive and limit the potential damage. But there will be nothing you can do to prevent attacks on unguarded civilian targets – this is out of your hands.

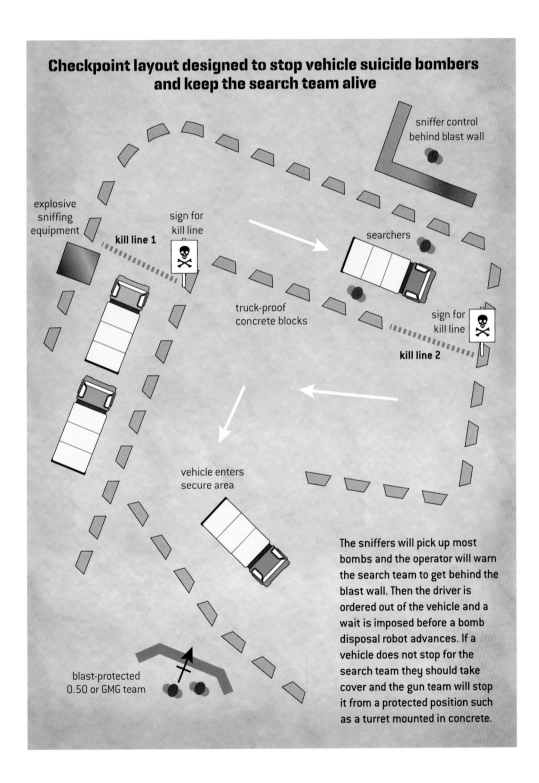

# Checkpoint layout designed to stop vehicle suicide bombers and keep the search team alive

sniffer control behind blast wall

explosive sniffing equipment

sign for kill line

**kill line 1**

searchers

truck-proof concrete blocks

sign for kill line

**kill line 2**

vehicle enters secure area

blast-protected 0.50 or GMG team

The sniffers will pick up most bombs and the operator will warn the search team to get behind the blast wall. Then the driver is ordered out of the vehicle and a wait is imposed before a bomb disposal robot advances. If a vehicle does not stop for the search team they should take cover and the gun team will stop it from a protected position such as a turret mounted in concrete.

When effective searches are either not implemented or followed through... The aftermath of a massive explosion which killed at least 18 people including children and security officials and injured more than 100 others in Karachi, Pakistan, on 11 November 2010. (Corbis)

large investment by the terrorists in a truck bomb both in terms of planning and logistics including supplying the explosive. If it does happen, try a head shot at the driver, it might just stop him pressing the button – what have you got to lose?

What is much more likely to happen is that a vehicle will be driven at a position or building you are guarding. What are you going to do then? By the time it happens it is too late to think about it. Defence against motorized bombs is all in the planning. So far as is possible, given the fact you may be in a town, prevent vehicles from approaching within blast radius of the target by placing large concrete blocks or pillars sunk deep in the ground. These should be sufficient to stop a heavy vehicle approaching at speed and create a perimeter all around the target – not just where the entrance points are.

At the point of entry force vehicles to slow right down on their approach by placing further large concrete blocks or similar. These should be set up in such a way that the vehicles not only have to slow down but they have to manoeuvre through 'S' bends at least twice **after being stopped** by the search crew. By this means, a vehicle is stopped and searched outside the perimeter where it cannot do severe damage to the target or most of the guards.

The search crew should always be covered by fire from a position within the perimeter. By this means, if the search crew are shot, or their commands to stop are ignored, the vehicle can be brought under sustained and accurate fire. Even if the vehicle cannot be stopped immediately the driver can be killed and the 'S' bend will stop the vehicle's closer approach.

As with walking bomb checks it is reasonable to draw for, and rotate, the job of search crew.

# CHAPTER 9

## Defending a Position

# INTRODUCING THE IDEA OF DEFENCE

In this section we are going to look at how you keep yourself and your mates alive when you are attacked and the enemy are doing their best to kill you. As we have already seen, a lot of defence comes down to digging a hole and sitting in the bottom of it — not very glamorous but comforting when the two-way range is getting busy.

First of all we will take a look at classical defence; how you pick a place to defend, how you design the defences and how you man them. This will give you a clear idea of the principles involved in a prepared defence so you can then go on to design your own as the circumstances require.

Then we have what we might call the unprepared defence. Situations where you are not specifically dug in and ready for the enemy but, being a good soldier, you always have it in the back of your mind that you may be hit at any time and have your SOPs ready. This covers defending yourself on a foot patrol or convoy against ambush. Then finally we look at manning checkpoints and the bane of every private soldier's life — sentry duty — where you are in a permanent condition of defending a particular target.

Its all pretty straightforward stuff but if you get it right you will stay alive and if you get it wrong you may not — so make your decision carefully regarding learning this part. Listen to the news and think how many of the casualties you hear of would still be drinking beer in the

A British Parachute Regiment patrol comes under threat and assumes an all-round defensive position. Notice there is a sniper with the patrol. (Photo courtesy Tom Blakey)

mess if they had done it right. As I'm finishing this section I have just heard the news from Afghanistan. A bunch of guys were caught in the accommodation area of their base by a couple of mortar shells. Some were hurt badly and one died because they didn't have a hole to jump into. If they had been operating correctly they would still be on their feet.

US Marine Corps base near Kandahar in 2001. Solid walls, strong roof, blinds behind to stop silhouette. Good effort. (Topfoto)

## Defining the threat

Wherever you are and whatever you are doing when attacked, the first thing you need to think about is what the enemy is trying to achieve. Yes, I know you are thinking 'That Bastard is trying to kill me,' but actually the reality is a little more complicated than that.

Many, indeed most, direct attacks by insurgents are not aimed at overwhelming your defences and killing you all. They are harassing attacks designed to make life unpleasant for you. But, of course, there is always the potential for a 'proper' attack where the opposition are out to slaughter every man Jack of you.

As a general rule, if you are in a prepared position, the insurgent attackers are out to make your life miserable because you are too strong to assault. But if you are on patrol and ambushed or in a tiny, remote position with no back up, you need to work out if the opposition is just harassing you or out to kill every single man in the team.

You need to assess which type of attack is coming in because what the enemy is trying to achieve will tell you a lot about how many of them there may be, what their plans are and what may come next. This Intel is always useful in a contact situation as it guides your response.

So how do you know what they are doing? Let's take a look at the two types of attack and what they are seeking to achieve.

### Harassing attack

As a rule, a hit-and-run harassing attack is the tactic employed by a force which does not have sufficient troops or weapons at its disposal to mount a more effective attack – an attack capable of causing more casualties or capturing the ground. The harassing attack may be mounted against a huge encampment by one or two men with a single mortar tube or a sniping rifle. And they have a good chance of getting away if you are not properly organized. We will see about that shortly.

The most common form the harassing attack takes is the occasional mortar round fired in the general direction of your base but intensity of fire can be high and weapons employed may include rocket launchers and small arms in the indirect fire role – i.e. fired in the air to drop on you.

The purposes of the harassing attack include:

- Preventing you taking rest
- Showing you and local civilians the presence of terrorist forces in the area
- Killing a few soldiers or civilians in a given area
- Lowering morale on the receiving side

### Assault to contact

Tacticians have always considered that an assault on a dug-in position requires a numerical superiority of 3 to 1 in favour of the attackers. This and either no defensive air operations or air supremacy in favour of the attackers. Basically because men in holes are harder to kill than the men walking across open ground towards them.

You will rarely experience an assault to contact against a prepared position in anti-insurgency warfare. Against you at any rate because as Special Forces operatives there may be times when you are acting as insurgents. The main reason for the rarity of insurgents assaulting prepared positions is that you will always have air superiority and if the insurgents got enough men together to mount an assault against a main base – or a prepared position of any strength – your air support and artillery would have a field day and destroy them utterly.

The only time you may be on the receiving end of such an assault is if you are dug in a remote location as part of a small unit. This is where you have to be careful with the preparation of your defences as we see in the next section.

Where the insurgent enemy do attempt an assault to contact their method will go something like this:

- Gather local superiority of firepower quietly and quickly
- Over-run your position quickly without a preparatory artillery barrage and kill everyone
- Gather what Intelligence, photographs and propaganda material presents itself
- Melt away before reinforcements or the air-cover arrives

**REMEMBER:**
Don't think you are safe because you have 250,000 troops 100 miles away. *Localized* superiority of firepower will always win a firefight.

# DESIGNING DEFENSIVE POSITIONS AND DIGGING A HOLE

In this section we are going to look first at how to dig yourself a nice, comfortable hole in which you can sit all safe and snug from shot and shell while shooting the attacking enemy. Then we will see how a group of these holes can be arranged in such a way that attacking them is more trouble to the enemy than it is worth. This is what is meant by a prepared defensive position.

As a soldier you are going to spend a lot of time in a hole in the ground or holding some God-forsaken mud fort waiting for the bandits to come and try to take it from you or otherwise spoil your day. Either they will be taking pot shots from the middle distance and lobbing mortars shells to harass you, or they are going to come over the hill by the thousand, screaming. In either situation being prepared is crucial and knowing how to set up a strong position is a useful and portable skill.

## The limitations of defensive warfare

You will not win a war by defence because the enemy will only come to you when he is strong and if you hurt him he will run away to lick his wounds. Despite this, over time you may weaken the enemy sufficiently for him to collapse and you will certainly achieve other strategic objectives such as controlling ground, supply lines and bases. More importantly, from your point of view, you stand a much better chance of survival sitting in your own defensive position rather than running towards someone else's with a bayonet fixed to the end of your rifle.

## Designing a hole

Sand, earth and stones all stop bullets and shrapnel very efficiently. The trick is to keep them between you and the incoming threat. Of course, enough steelplate and concrete do too but the advantage of dirt is that you don't have to take it with you – it's just lying around everywhere. When you are dug in to the ground like a rat, with a roof supporting 18in of dirt, you are well protected from the direct fire of small arms and RPGs until you stick up your head to shoot back. You are also totally protected from mortars and shells coming in from above.

The author lived in this hole for nearly a month. There was either a mortar attack or direct assault most nights. The choice was made to dig in under the trees to stop the sun and risk an air burst from the mortars. Good decision as it turned out. (Author's Collection)

You are safer than a lot of places even if you don't have a mortar resistant roof. Firstly because there may be no mortars used against you and secondly because the chances of a mortar round dropping right into your hole are absolutely tiny and a near miss will not hurt at all.

As you squint over the edge of your hole the only thing the enemy can see to shoot at is the whites of your eyes. And if they get that close you are doing something wrong. The only way anyone could realistically hurt you is with heavy artillery, 155mm and upwards, firing ground penetrating shells or special bombs which do the same thing. Insurgents generally don't have this kit.

At a push, the minimum you need to be safe from shot and shell is a hole which gets your butt just below the level of the surrounding soil. That doesn't have to be a very deep hole. Which is just as well if you have to dig a new one every night with your own fair hands and a government issue entrenching tool – and the ground is like a concrete runway.

If you are just stopping for the night, and there is enough of a threat to warrant protecting yourself by digging in, then the smallest hole you can sleep in with no 'fancy work' is just fine. The level of threat requiring a hole can loosely be defined as the enemy knowing where you are and having the wherewithal to take pot-shots at you. So that means *always* if you are in an established base-camp and *usually* when you are out on patrol. After all you can't often be certain you weren't followed and that no one unfriendly will trip over you in the night.

Anything more than a night or two and you may want to make yourself more comfortable. The smaller the hole the less digging but a smaller hole is much less comfortable. On a long stay your hole is your home and you want all the mod cons. Any fool can be uncomfortable.

> **REMEMBER:**
> Leave your kit on the surface next to your hole and you can rely on it getting peppered with shrapnel or bullet holes.

To be comfortable the least you need is a hole big enough to stretch out to sleep, together with all your kit. It is also nice if you can sit up and make a brew while you are 'standing to', being shelled or whatever, so it's a bonus if you can get it deep enough to keep you safe while you sit upright. A little over 6ft long by about 2ft wide is enough for one man but it is much more pleasant to build two-man holes for company and to keep each other awake. That would be more like 8ft by 3ft and then you can lay with your heads at each end.

Dig your hole 3ft deep and you are making some serious work for yourself using an entrenching tool in hard ground. If you have not seen one, an entrenching tool is a cross between a small shovel and a pick-axe which folds up for carrying. It is easy to carry but not the best possible digging tool. Digging a hole 8 x 3 x 3ft means loosening and moving $2\frac{2}{3}$ cubic yards of whatever is underfoot. If you are stood on earth then packed earth weighs about one and a half tons per cubic yard. Which is plenty.

If time permits, an elbow-shelf along the front of the hole is excellent for comfort while holding the endless firing position and also stacking magazines ready for use, water and other kit. On a long stay you might have the time and energy to put a roof on your hole. Between 12 and 18in of dirt will stop any mortar and a plastic sheet under it will keep the rain off. A roof is a definite benefit if the sun is very warm or there is constant rain. A poncho can be used to protect against the elements but they are not so good against mortar bombs. When you have a roof, always try to drape something down to the rear so that your head is in shadow and does not make a good silhouette target for the enemy.

## Digging technique

The ideal way to build a base is to have the engineers come with heavy plant machinery and either digs holes for you or fill those lovely big bags they have with sand. Failing this, have one man loosen the earth with a pick and the other scoop it out with a shovel. This way is much easier and quicker than any alternative. Try never to break packed earth with a shovel or spade.

If you can get hold of a bunch of sandbags then you can cut down on the amount of muck you have to move. If you dig down 18in then you have sufficient earth to fill enough sandbags to build a wall 18in high around your hole. This not only makes for less work but gives you a raised shooting platform with a better view of approaching enemy creeping along the ground.

If there is constant heavy rain your hole is going to get wet and muddy unless it is in sand. There is nothing you can do apart from dig a 'sump' and keep baling it out. Constant rain is like Chinese water torture and really gets you down after a while — trust me the effect of the constant beating of rain on your head has to be experienced to be truly believed.

## Sandbags

Ordinary man-portable sandbags about 18in long are very useful for constructing a bullet proof wall out of next to nothing. A dozen empty bags weigh ounces but filled with 'muck' a sandbag placed lengthwise will stop a rifle bullet at point blank range. This is because the sand or earth spreads the force of the strike outwards from the point of impact in a cone and it is dissipated in moving a lot of sand a little way. With a little imagination you can find all sorts of ways to improve your condition by the use of sandbags. You can also create little walls between foxholes to allow movement under cover without endless digging.

A raised shooting platform which allows you a better shot at the enemy than firing from ground level is a good idea too. It also reduces the chance of bullets bouncing up at your face from the floor. If you do this remember that a sandbag wall is a magnet for the RPG rocket. Because the blast from this weapon travels forward with tremendous force one layer of bags will not save you. More likely you will be sand-blasted to mush and cooked at the same time. The way to proof a wall against RPGs is to build a single wall outer layer to set off the charge and a double inner layer to absorb the force of it.

While single-man holes are the safest and most efficient in terms of spreading out your forces to cover a perimeter and minimizing loss of firepower from casualties, it is a fact that many men 'soldier' better with their mates by their side. If circumstances permit you might want to consider allowing double or group holes to allow for a communal smoke or game of cards in between the shooting matches.

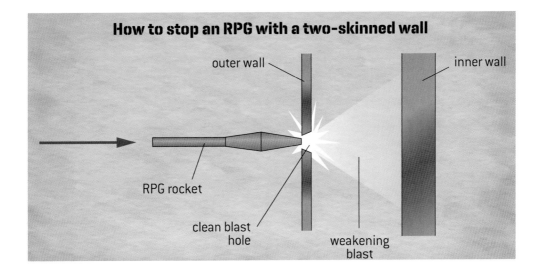

**How to stop an RPG with a two-skinned wall**

outer wall

inner wall

RPG rocket

clean blast hole

weakening blast

## Trenches

A trench is only a long hole but it has several features not at first apparent: on the positive side it allows the movement of firepower to where it is needed without exposure to enemy fire. It also allows good command and control – and the sight of officers and friends is good for holding the men together in a warm action.

Khe Sanh, a remote US outpost in Vietnam, faced full-scale siege from the North Vietnamese forces during the Vietnam War until it was finally abandoned two months after the siege began. Someone there knew how to dig trenches. March, 1968. (Corbis)

On the down-side an enemy standing at the end of a trench can kill everyone in it with a burst of fire or a few grenades. One mortar hit is likewise destructive. The classic way around this problem is to dig the trenches in a zigzag pattern. It works but it is a lot of digging.

## Gun pits

A machine-gun hole needs to be a little larger and will benefit greatly from a raised firing position if the situation and manpower permits. As it is a focus for any serious assault you might want to consider a roof to stop grenades or aimed mortars.

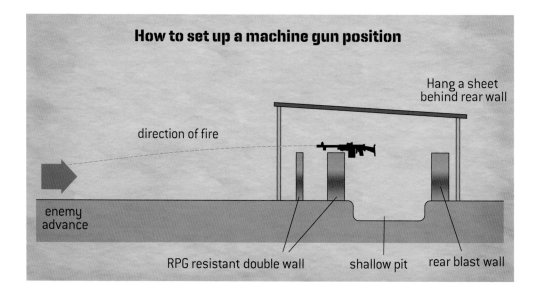

**How to set up a machine gun position**

Hang a sheet behind rear wall

direction of fire

enemy advance

RPG resistant double wall | shallow pit | rear blast wall

## HQ and support bunkers

With a unit larger than, say, platoon size (36 men) it becomes useful to centralize command, signals, medical aid, ammunition supply and ration stores. Each of these can have its own bunker which should be large enough to accommodate the relevant personnel as well as their materiel. Because each of these groups are of great importance to the whole formation it is usually sensible to build them a bomb-proof roof wherever possible. As it is already becoming clear on operations you will always be working either as a donkey or a mole.

## TOP TIP!

### Alternatives to digging

There are three common alternatives to digging holes yourself: you can get the engineers to dig the holes with machinery, use explosives if you are lucky enough to have them on hand – or failing one of these blessings you can persuade the locals to dig the holes for you. A food for work programme often works well.

Getting the locals to dig holes is the most common and preferred alternative of the seasoned infantryman as you can bet your last dollar the locals will happily dig you an excellent hole for a can of something meat-related: and then they will immediately trot off to tell the local insurgents where you are. Either because they hate you or they don't want the insurgents to cut their bits off. In any event, this saves you endless patrolling and sore feet with the added benefit that you are fighting from a prepared position chosen by yourselves at leisure. I have used this trick countless times and it has never failed.

## Buildings

You may well have to hold buildings against attack. The problem is that roofs create air burst over your head from mortars and heavy machine guns and RPGs go straight through most brick walls. The reason is that brick is a brittle material and doesn't spread the shock of impact all that well compared to packed sand or gravel. Your best bet is to build a second skin to the walls you intend to shelter behind out of sandbags so as to trigger and suppress RPG rockets and provide some protection against heavy machine guns.

For sure a heavier type of weapon such as a cannon shell or a machine gun of greater than rifle calibre will make short work of a wall whatever you do to reinforce it. The best thing is to dig holes for shelter and fire positions and use the building for cooking.

## Towers

Some permanent bases have towers already built so presumably they were built properly. With a tower you get a great view of the enemy from a comfortable, dry position and should

Concrete watch towers with wire surround to prevent RPG7 damage by detonating them too early. I would prefer that wire to extend above the top of the tower. Camp Bastion, Afghanistan, September 2010. (Corbis)

be able to fire onto anyone creeping towards your position. This is a blessing. Less good is the ability of an RPG or .50cal machine gun to punch holes in anything but the stoutest tower.

# DEFENSIVE POSITIONS: SELECTION AND LAYOUT

As every defensive position is different owing to the ground, weapons available, defending strength and so on, I am going to show you what to look for in selecting a defensive position and how to lay out the perfect defence. You can then mix and match the principles to cover the less-than-perfect circumstances in which you will doubtless find yourself.

At the root of the layout issue are the two conflicting requirements of a defensive position: these are to be spread out as much as possible against mortars or artillery and to be as small as possible so your forces and weapons are spread around as short a perimeter as possible. It doesn't take an 'Einstein' to see that the shorter the perimeter you are defending the closer your guns are together and therefore there more concentrated the fire which an attacking force has to endure.

First of all here is a list of natural features you want to take into account when selecting your stronghold:

■ **Height:** Wherever possible set up a defence on a hill. The reason for this is that if you overlook the surrounding area you can see the enemy advancing and you can see your fall of shot to correct for accuracy. Of course, avoiding having the enemy overlooking you also stops him correcting his own fall of shot. This is not a new idea.

And it is also much harder work skirmishing to the attack up a hill. Not surprisingly this is why castles are mostly on hills. Being on a hill or at least raised ground also lessens the chances of being flooded by a downpour. Not fatal but unpleasant enough.

■ **Trees:** Avoid tree cover when you can, even when this means giving up the shade, because mortars and artillery falling onto trees become air burst. See *Not Getting Shelled – High Explosive*. Trees and bushes around your base should be removed as they let the enemy get close before you see them.

■ **Ground:** Pick ground you can dig a hole in if possible. Digging into baked or rocky ground can not only be demoralizing but it can encourage soldiers to 'forget' to dig in properly. Avoid land which may get flooded if you can.

■ **Insects:** Some places have the most annoying insects like wasps, ants and various types of ticks. So far as you can, check out your site for these before settling in.

■ **Access:** With a larger base you are going to want ready access for vehicles to re-supply if possible. If there is only one road you may find it mined and ambushed so often it is unusable. It is always best to have a range of possible exit points so as to allow you to get patrols out and back in without the enemy knowing where they can catch you. Consider also the possible need for re-supply by fixed-wing aircraft, helicopters and boats.

You may also want a route out of the base which is sheltered from fire to escape if this becomes necessary. Weigh this against the possibility of offering cover to approaching enemy forces.

## Spreading the risk

We have already seen that the shorter your perimeter the more concentrated your fire can be at all points along it.

What is not quite so obvious is the advantage of spreading your forces over an area. Someone shelling the position in the example to the right with mortar fire from a safe — for them — distance away has a beaten zone — the area where his bombs are going to fall — of roughly 300ft x 300ft = 90,000 square feet. A bomb would actually have to land in your trench to do you harm and your trench is 6ft by 2ft = 12 square feet.

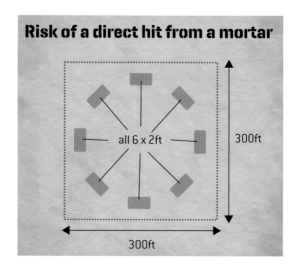

**Risk of a direct hit from a mortar**

all 6 x 2ft

300ft

300ft

So the maths say there is only 1 chance in 7,500 of any individual bomb landing in your trench. That is better odds than the guy firing the mortar has. It's better odds than crossing the highway safely in some towns.

## Killing ground

All around your position you want a clear area of as large as possible. 100 yards minimum by choice. 300 yards is better. The two reasons for this are:

- ■ To stop the enemy shooting at you from close-up cover
- ■ To make the enemy run across a lot of open ground in his final assault to contact

If there have to be any ditches or hollows in the killing ground which you cannot see into, and which therefore may give cover to an attacker, fill them with mines and razor wire. Zero-in mortars, artillery and fixed-line machine-gun fire if you are a larger unit.

## Laying out the position

Once the site of the position is selected one person should be authorized to mark out all the defensive fire positions. A circle gives the shortest perimeter for the greatest area covered.

Normally two men to a foxhole is good as single foxholes can get lonely. Also, if one man is killed its whole arc of fire is not left undefended.

Arcs of fire are allocated to each position and these arcs should not only interlock but should still interlock when one position is taken out of the scheme.

An 'arc of fire' is the angle of responsibility given to any man or group to cover and defend. In the diagram below it can be seen that not only is the position unapproachable without coming under fire but it will remain so even if any of the foxholes are silenced.

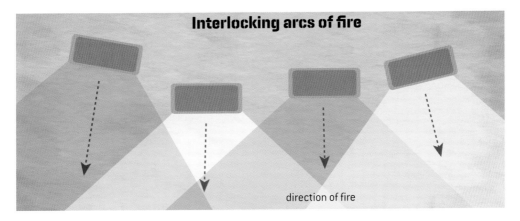

**Interlocking arcs of fire**

direction of fire

Machine guns should be sited at points of particular danger – such as where the enemy can get closest to the position without being seen or fired upon. Snipers should be positioned where they can see the furthest or are otherwise of most use.

As I said earlier, if you are going to be staying in a position for some time it can be a good idea to dig communicating trenches between the fire positions so as to allow reinforcements to be moved around without exposure to fire. With a larger unit you might want to consider building a command post and centralizing food, water and ammunition supplies. There might also be room to establish a first-aid post – medics are always good for morale.

## Wire

Military barbed-wire is like the stuff farmers use but the spines are longer and closer together. Razor-wire is a central strand like barb-wire but with blades like old fashioned razor-blades in place of the spines. If you are in a hurry to cross a wired-up defence you can tolerate rolling over barbed wire but razor wire is much less fun. Both types come in coils which are opened out into loose spirals when in position.

Wire doesn't stop soldiers it merely slows them down so you have more time to shoot them. Attackers tend to get caught up in wire and have to disentangle themselves. It is best set loosely in coils and tangles because this way it is harder to clear a path through it and, when entangled, harder to get free. Wire should always be covered by fire. Preferably machine-gun fire. And a few mines are no bad thing either. There is no point slowing down his advance if the enemy can safely take his time.

### Landmines

Landmines – particularly anti-personnel mines – are pretty unpleasant things: but no one said it was going to be all fun and games in the army. Use landmines to make the enemy reluctant to approach your position if you are staying for long enough to justify the effort. When an attacker treads on one of your mines it will seriously undermine the morale of his comrades as he will likely lay around screaming for hours. Or until you shoot him as all defensive minefields should be covered by fire.

### Booby traps

It is often a good idea to set booby traps such as trip wires attached to flares or Claymore Mines in places like gulleys where the enemy are likely to approach your position. By careful use of this technique you will often catch the reconnaissance party coming to have a look at your position. My own preference is to arm the traps last thing at night and disarm them first thing in the morning detailing the clearing patrol for this duty.

## Response to harassing fire

When you are properly dug in your main irritant will be harassing fire from mortars and the odd RPG lobbed inside your position. There are a number of things you can do about this besides making sure you are in a nice safe hole.

The low-tech answer is to watch out with night sights and spot where they are shooting from. A mortar or RPG is not hard to spot at night if they are within visual range – and an RPG has to be to have a hope of hitting you. A mortar will probably be in sight unless they are well set up and you can certainly see the flash easily enough at night.

If you can get the kit there are two other actions you can take: you can fire back accurately at the mortar launch point and you can shoot down the bombs in mid air. Yes, really!

## Artillery Locating Radar

Since Vietnam there has been a radar system which picked up and tracked incoming mortar bombs, shells and ballistic missiles.

The radar takes a number of snap shots of where the projectile is in its flight path and calculates from these where it must have been fired from. Then it either prints out a grid reference for the firing point or in the later models it lays your artillery onto the firing point and takes out the people annoying you.

The Danes have developed a neat system called MAN 27.314 DFAEC ARTHUR which is mounted on a truck which you just park in your base. It spots any bombs or shells coming towards you and works out where they will land, where they have come from and lays your guns onto their firing position. It is efficient out to a range of about 40km or 25 miles and can handle eight targets at the same time and about 100 targets a minute. Which is pretty good I think. This system has worked well in Afghanistan for the British and other nations have their own variants. Of course you won't get the use of this unless you are a big unit.

The skilled insurgent will fire a few mortar rounds and move to avoid the above response but just the threat of retaliation limits the time of attack and therefore the damage that can be done to you and your position.

## Counter-Rocket, Artillery and Mortar (C-RAM)

Just in case you are timid and concerned about a few mortar rounds landing in your base before the counter-artillery radar wipes the firing party out, there is a novel protective system becoming available which uses a radar to spot and track incoming rockets, artillery shells and mortar bombs similar to the Artillery Locating Radar – then it lays on its own 20mm M61 Vulcan Gatling gun autocannon, and shoots them out of the air. I kid you not. It was developed by our friends in the US and called the Phalanx officially and R2D2 unofficially owing to its likeness to a Star Wars droid character. Don't you just love it?

The British have had their own version, designed and built by the Dutch, which is deployed on all British warships of any size and is called Goalkeeper, as a football/soccer analogy. Similar in operation to the Phalanx it has a 30mm roller cannon and opens up with 70 rounds a second until the target is eliminated. This has been tested for air base defence but it is heavier than Phalanx and therefore not as portable.

### Reconnaissance drones

Another defence option which will become more available as technology spreads its influence is a small reconnaissance drone which you can launch and use to monitor the opposition from above. They can even be fitted with heat detection and other clever stuff which I had better not mention here but which allows them to spot snipers, RPGs and mortars very easily in daylight or darkness. Happily for you, if the insurgent is in dead ground and leaving after firing his rounds the drone can still spot him. The drone will just transmit a grid reference to you or your artillery support and you can shell him to jelly before he has walked up a sweat.

### Air assets

If a terrorist is foolish enough to maintain his position for a while then you can call in an air-strike. For those who have never experienced an efficient air-strike it is like you really got on the wrong side of God. There is no answer. The whole area is covered in cannon fire, shrapnel or napalm and it is difficult to do a 'body count' for the mush and bits lying around.

### Snipers

But what about snipers in daylight when you don't have any clever kit? You already know they are extremely difficult to spot. Probably the best you can do without special equipment is get the general direction from the thump of the rifle fire and gauge the rough distance by the time difference between the crack of the bullet passing overhead and the thump of the shot reaching you. For the hard of thinking, a rifle bullet travels faster than sound so it reaches you quicker than the sound from the rifle muzzle so you hear the 'crack' from the sonic boom before you hear the 'thump'. The further apart these two sounds the further away the sniper. At ranges over a couple of hundred yards it starts to get noticeable and at long range very much so. A simple answer to an irritating sniper is to set up a dummy in an exposed position with a hollow body and some stiff card front and back. The idea is that when a sniper puts a bullet through the dummy you only have to look through the holes to see where it came from. Then you can use your own sniper, artillery or a rocket to take him out.

## Response to a committed assault

The key point about defending against a properly organized assault is to maintain good command and control from a central position. By doing this you can ensure you know how

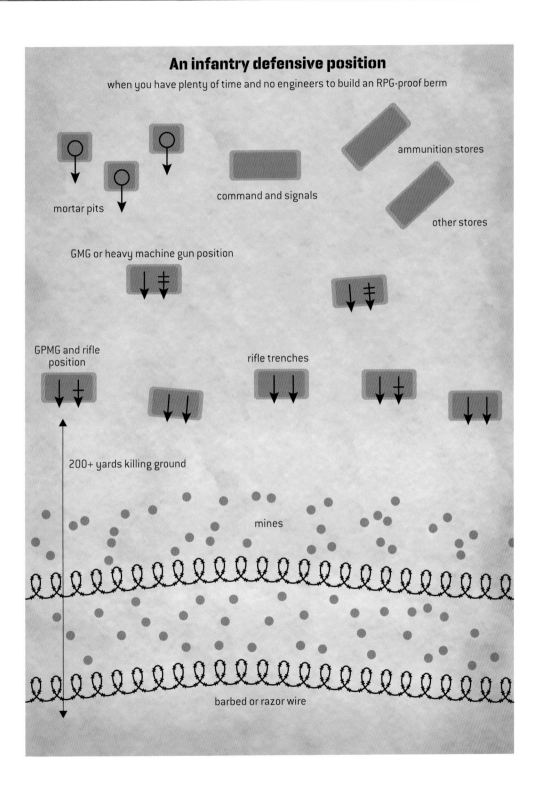

# An infantry defensive position

when you have plenty of time and no engineers to build an RPG-proof berm

ammunition stores

command and signals

mortar pits

other stores

GMG or heavy machine gun position

GPMG and rifle
position

rifle trenches

200+ yards killing ground

mines

barbed or razor wire

the threat is developing and move your forces appropriately to strengthen the perimeter where it is under pressure or call in air strikes against specific threats.

You would be well aware if a classical attack was on the way as there would be a prelude including a thorough stonking by an artillery barrage to soften up your defence and organization. This will not happen in counter-insurgency as your artillery or air assets would destroy their artillery utterly in a few minutes and be glad of the chance.

What you might get in an insurgency operation is a surprise attack with the intention of quickly overrunning your position. This is how it might go: all of a sudden there is heavy incoming fire from machine guns and RPGs. Within moments a line of insurgents come racing across the open ground towards your position hoping to take advantage of you keeping your heads down.

It is up to you not to hide in the bottom of your trench but to return heavy, accurate fire. Your unit commander should be bringing fire to bear against the enemy suppressing fire and while he does that you need to stop the enemy closing with your defensive line.

If you are set up properly the assault group will be advancing over open ground, uphill across razor wire so they are not coming too quickly to shoot down. As they must be in the open to advance you can bring mortar or artillery fire to bear and pretty much any attack will be neutralized.

Does this sound too easy? Too simple? If it is easy, is it easy because you set up the position properly well before the attack kicked off. Your radios all worked well. You had your men in safe positions and your mortars in the middle of your base were ranged in on likely points of attack. Your wire was spread thickly across the killing ground and your guns covered it all. Train hard fight easy.

# ANTI-AMBUSH DRILLS: PATROLS AND CONVOYS

A correctly executed ambush should be just that – execution. Everyone in the killing zone should be dead within 2 or 3 seconds. We will see in the next section how to set up an ambush properly and achieve that effect on the enemy, but here I want to go through the Standard Operating Procedures first of all to avoid being caught in an ambush and then for when you are ambushed yourself.

Given the ambush you are caught in is not set up properly then react correctly and you will have a good chance of survival – and you may even nail the ambushers. If the ambush is set up well then the SOPs give you something to think about for a few seconds...

Don't get over excited about being ambushed; I have been ambushed scores of times and never has an ambush been set up or carried out properly. Obviously, as I am sat here rattling my gums.

Anti-ambush drills are for when you have been unable to avoid being ambushed: the ambush has been sprung and people are shooting at you. They work when the ambush is either poorly organized or the intention is more to harass your people and make you run around than kill you all. This latter is often the case when insurgents don't have the manpower or hardware to put on a proper show. Anti-ambush reaction drills do not work well when the ambush is set up right. Nothing does.

## Avoidance is better than reaction

Whatever your strength and assets the best defence against an ambush is not to be caught in one because, generally, an ambush is sprung at the choice of the ambusher. This means the ambush is only triggered when the target force is either weak enough to kill or unable to chase down the group applying the harassing fire. You will always be at a disadvantage when ambushed. Your first consideration then is to organize your forces and movement to avoid an ambush. If your anti-ambush patrols catch an enemy force preparing an ambush for you then they are effectively assaulting them or ambushing them first. This is much better.

The basic principles of getting a supply convoy safely from Point A to Point B haven't fundamentally changed in 40 years. A supply convoy in Yemen during the counter-insurgency campaign in 1964. (Photo courtesy Colour Sergeant Trevor 'Sadie' Sadler, 1st Battalion 'The Vikings' Royal Anglian Regiment)

A British SF patrol in Jackals, Afghanistan 2010. (Photo courtesy Tom Blakey)

Always put a steady man on the back vehicle: former Para, Sergeant Roy Mobsby, PMC vehicle patrol in Iraq. (Photo courtesy Roy Mobsby)

Given you don't come across the ambush before it is sprung then how you respond depends to a great extent on your forces and the type of ambush. I have tried to cover all the alternatives below to make this clear and get it to stick in your mind.

## An ambush against your foot patrol

- **Avoidance:** As a foot patrol you are not obliged to stay on roads and this freedom makes your movement more difficult to predict and therefore trickier to ambush. The ideal way to protect against ambush of a foot patrol is to avoid obvious tracks and have a party moving ahead on either side of your main force. This means they come across any ambushers from the ambushers' flank and at worst give you warning, at best, shoot a few. See diagram on page 276. Of course, sometimes the ground does not allow this option or you are too short of manpower etc. In which case you will get hit now and again.
- **Reaction:** As a foot patrol moving in enemy territory, or at least country where the insurgents have a presence, you will be ready to fight as opposed to being burdened like pack mules with supplies. Usually the ambushers will hit you from too far away to do the job properly. By that I mean kill you all. An ambush where fire is opened from a distance can be dealt with as for the patrol 'coming under fire'. You take cover, assess the situation then organize an assault on the enemy or call in an air strike. By far the greater part of the time, ambushers in an insurgency situation open fire from a distance because they lack nerve, tactical knowledge or just want to run you ragged. Beware the one time when they are being clever. A few rounds fired, as if from a badly set up ambush, can be used to draw you into the perfect ambush position.

A close-range ambush is more serious as the choice to run away should have been eliminated in the planning by mining or covering the escape with effective fire. The only way to go is straight at the enemy. A determined charge, perhaps after setting up a fire team on your flank, will cause most ambushers to lose their nerve and run.

**REMEMBER:**

The best anti-ambush drill by far is to avoid being caught in one.

Certainly this is better than the alternative – stay where you are in the area chosen by the enemy as the easiest killing ground.

## An ambush against your vehicle patrol

Pretty much the same applies as for foot patrols with one or two significant differences:

- **Avoidance:** Vehicle patrols are more likely to stick to roads or at least driveable areas and this makes their movements more predictable. Even in open country, between two points there will often be bottlenecks creating an ambush target. Because vehicles move so quickly, it is probably impossible to have ground troops ahead on either side of the route running anti-ambush, so air cover is the only answer. A drone will do the job nicely as it can stay in the air for hours, move quickly enough to keep up with the vehicles, its sensors can spot hidden enemy by their thermal signature, and it can easily be operated from within a vehicle. Spot the ambushers from high in the air and they will not know it so you can seriously turn the tables on them.

    Vehicle patrols tend to carry heavier weapons than foot patrols and be armoured to some extent. This means that an ambush will almost always be triggered by a mine or IED and then, rather than make the effort to kill you, the ambushers will very often fire with RPGs or heavy machine guns from a distance in the hope of penetrating your armour and increasing the casualties without too much risk. An RPG has a range of 1,100m and a .50cal machine gun double that so very often a small ambush party will stop you with a bomb and then fire at you with a heavy machine gun from over a kilometre away up a seriously steep hill. The best way to avoid a vehicle ambush other than the use of drones is anti-IED measures.

- **Reaction:** Any kind of ground assault reaction to an attack from 1,000 feet up a hill is a laughable waste of energy. I can testify to this as I have been sent up hills against ambushers. By far the best response is to use the heavy weapons on the vehicles to make the position of the ambushers uncomfortable and continue on your way. A 30mm cannon will cut through most built defences but better is something heavier like missiles, mortars or 90mm guns as then the shrapnel will penetrate nearly all defences. Of course, best of all, is a nice air strike.

## Protecting a supply convoy

Where you totally control an area there is going to be safe passage for your troops and vehicles. Where the enemy control the area you aren't going to be sending supply convoys through, you are going to be attacking the enemy. The time that a protected convoy is used, and is at risk, is when you are occupying an area but the enemy are still present in sufficient numbers to potentially mount an ambush.

By its very nature a convoy of trucks is something of an easy target for an insurgent attack. Trucks have to follow recognized roads and sooner or later you have to move your people, equipment or whatever. Moving everything by air is expensive, slow to move bulk and ties up air-assets – in the end you really need to use trucks even though they are not heavily armoured. So – in principle – all the insurgent has to do is sit by the road.

In addition, following a road – which may wind through difficult and mountainous country – means the convoy is difficult to protect. It is a thin line of firepower at best and the enemy can concentrate his forces at a point ideally suited to his needs. This makes it relatively easy, in the ordinary way of things, to create a localized superiority of firepower in favour of the attacker.

Another advantage to the attacker is that he has as long as it takes to make his move – where there is a threat of ambush you have to protect every convoy tying up valuable resources permanently while the terrorist might not even be in the vicinity. You might send a convoy along a road every week for a year, tying up all manner of defensive assets, and never be attacked once. Or just the once. But you never know – this is the constant advantage of the insurgent position.

So what can you do to keep your convoys safe? Well, as ever, the answer depends on several factors which interrelate. Chief amongst them are the assets at your disposal, the assets at the disposal of the enemy and the type of country you are driving through.

### Friendly assets

**Aircraft:** Helicopters or rapid response fixed-wing aircraft are the most effective type of protection for a convoy. It is not that they prevent an attack, rather that they ensure the enemy are killed and don't do it again. With modern facilities for target acquisition, such as infra-red cameras, it is difficult for the enemy to escape unless they are in very mountainous or heavily wooded country. You should avoid taking a convoy through such areas when you can. But, of course, you often don't have a choice.

If choppers fly along the course of the convoy, clearing the ground in advance, the enemy are going to be picked up before they make their move or, failing that, as soon as they open fire. At this point the choppers bring overwhelming firepower to bear and the attackers are liquidated. With fixed wing you get a similar outcome but the aircraft would be on call and strike the enemy once they had been detected by the sensor aircraft.

The trouble is that a convoy can be on the road for days at a time and to be of any use the aircraft have to be available round the clock, or at least when the convoy goes through trouble spots. Then there may be a hundred convoys without a hit but all those assets are tied up. The terrorist wins without firing a shot or even being there.

Foot patrol through an Afghan village 2010 – waiting to be ambushed. (Photo courtesy Tom Blakey)

So, because of the cost in assets employed, aircraft are usually not available to guard your convoy unless the enemy are very likely to attack and the ground is particularly difficult. In this case an effort is usually made to move as much as possible in one go – a very big convoy.

In their war with Afghanistan (1979–89) the Russians had exactly this problem. The only way to protect their convoys as they travelled through the mountain passes was to use helicopters. Someone supplied anti-aircraft missiles to the Afghans and this made life difficult for the Russians. They lost a great many men and helicopters. Actually, the Americans shipped the Stinger missiles to Pakistan from where they were trucked to the border and a Dutch friend of mine took them into Afghanistan on the backs of camels. Like Laurence of Arabia with tulips.

**Ground forces:** When you are travelling in convoy through open country a useful way to prevent a surprise ambush is the 'Leading V' formation, also employed in infantry manoeuvres and mentioned earlier. In this case it can consist of infantry or even armour.

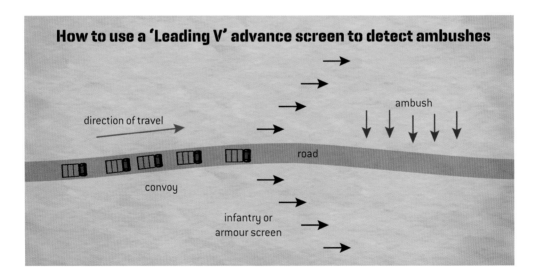

The principle of the Leading V is that the screen catches the ambushers in the flank, at their point of weakness, rather than providing the target to their front which they are expecting.

Where the ground makes a Leading V impossible, a useful second best is to send patrols out ahead to take possession of peaks and outcrops thus giving them the advantage of height and position over potential ambushers.

## Unfriendly assets

Almost any weapon can be used to ambush you but each has its own features, advantages and disadvantages to both you and the user. Direct fire weapons such as small arms and RPGs require the attackers to get up close and personal. This reduces their chances of escape when they are pursued by a stronger force. Indirect fire weapons are less accurate and can be eliminated by friendly air power as they are traceable and slow to move. Command detonated munitions – IEDs – are always very unpleasant.

**Country:** Wherever possible you are going to plan your route so as to avoid likely or useful ambush positions. A long drive is better than going through a narrow pass in the mountains. Knowing the principles of ambush yourself, and you will after you have waded through the next section, you can read the ground and assess where an ambush would be most likely to be successful or where the attackers might have a good chance of escape. The main things to watch out for are close cover by the road, high ground or difficult terrain on the 'attack side' and an open 'killing ground' opposite to prevent your escape to cover.

**Mines:** Where there is a ditch or killing ground opposite an ambush it is likely to be mined so be aware and think before you jump in for cover.

## Order of march

Travel well spaced to avoid both indirect fire from artillery and having all your vehicles come under fire from the same ambush party. If the trucks are well spaced a shell will only get one of them at worst and most likely it will miss altogether.

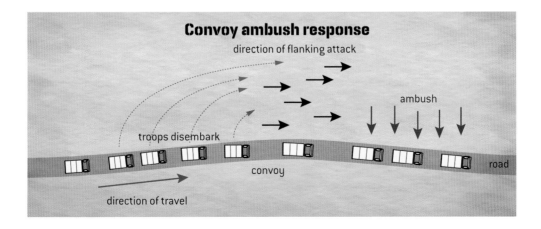

**Convoy ambush response**

A well spread convoy is also a threat to the ambushers because when one part comes under fire the troops in the remaining vehicles may be able to dismount and attack the ambushers in the flank as per the drawing above. All vehicles must keep in contact by radio and line of sight. The rear-guard must keep a watch for stragglers and breakdowns. The mechanics must travel at the rear in order to repair or destroy breakdowns as appropriate.

> **REMEMBER:**
>
> A favourite tactic of the experienced ambusher is to immobilize the front and rear vehicles first to trap the remainder and cause them to bunch up. So whenever possible, a vehicle capable of clearing obstructions, including your own vehicles hit by fire, should travel near the front of the convoy.

# CHECKPOINT AND SENTRY DUTY

These two are every soldier's least favourite jobs so don't think I haven't hated every minute of them myself. Checking thousands of civilians through gates in the wire or night after night of vehicle checkpoints in the rain... You know the score. The thing is, manning checkpoints and doing sentry duty are in many ways amongst the most important jobs you can do. You are keeping people, very often your own mates, safe from attack by bullet and bomb. It has got to be done so let's make sure we do it right shall we? And don't think the SF beret will let you escape these duties either – you will get higher value targets to protect by checkpoint and in a small team the guard rotas are harsh.

## Checkpoints

A checkpoint is usually set up static to either act as a deterrent and stop insurgents bringing bombs or guns into a secure area or mobile to catch insurgent operators and their equipment as they are moved around. In the following explanations I have ignored the technical issue of suicide bombers and the need to keep these at a distance as we covered this earlier. You will have to strike a balance in each theatre of operations between operational convenience and suicide bomber protection depending on the likelihood.

**Static checkpoint:** This is normally a permanent checkpoint on pedestrian or vehicle gateways in a perimeter fence or building entrance and manned by foot-soldiers. Your task here is to prevent the opposition bringing bombs, weapons or certain people into a certain secure area. Because the insurgents know it is there, the chances are they will not attempt to get through – providing you are seen to be doing your job carefully. All the people within the secure building or area are depending on you for their lives.

**Permanent pedestrian checkpoints:** When manning the access point to a building or area you know you are looking for a needle in a haystack and that most likely you will not find anything. Boredom and complacency are your enemies here as the cunning insurgent will know when you have been on duty for hours or your tour of duty has run for months. This is when he – or she – will strike.

What equipment you will be issued with to do the searching will depend on the circumstances and the priority of the target. Hopefully you will be issued with sniffer, electronic metal detectors or X-ray machines. Failing this you will have to make do with body searches. The important thing is that you check every person who comes through your gate. The reason for this is quite simple: by checking 1 in 3 or 1 in 10 people you make it risky for the insurgents to send in material. But not risky enough to stop them. Why would they care about getting one courier caught when they are prepared to blow themselves up? No, given the choice, you must check every person who comes to the gate and make sure you catch

**LEFT** A lone sentry with a machine gun stares at empty landscape with just a small goat for company, Afghanistan 2010. (Photo courtesy Tom Blakey)

**ABOVE** The rest of the Parachute Regiment patrol take a well earned break. (Photo courtesy Tom Blakey)

anyone who turns and does a runner. This makes the checkpoint an efficient deterrent and keeps both you and the secure area safe.

In some situations we are told it is impractical to check everyone owing to the volume of people. This is both dangerous and a waste of time as someone with a bomb will be sent to take the risk. But we all have to follow orders so be aware that a terrorist is not going to look like a shifty bomb-thrower. Rather, the most ordinary of characters. Make your selection for searching accordingly.

I suggest a random selection of prospects to search as a terrorist group testing your defences will observe your technique for some time before sending in their courier. If you don't search pregnant women, guess who will be carrying the bomb?

Another trick used by the terrorist is to send suspicious looking characters through immediately before the courier. This way you stop the suspicious person and are occupied when the courier carries their bomb through. Look carefully at the people following anyone strange-looking.

**Permanent vehicle checkpoints:** The three things about vehicles which make them dangerous to you are:

- Vehicles have a lot of places to hide things
- Vehicles can carry a lot of explosive
- Vehicles can force their way through weak roadblocks

Without using the boot/trunk, bonnet/hood or passenger compartment of the vehicle, you can still pack a huge charge out of sight. This means that a quick search of these places is not enough. Favourites spots are the spare wheel and the spaces inside the upholstery and bodywork. An unpleasant cargo such as animal dung is a good place to hide a bomb too.

In my opinion it is vital to have sniffer technology of some kind when searching vehicles for weapons or explosives. A dog or machine which can detect explosives and gun-oil/cordite makes life simpler, safer and allows you to search much quicker. Therefore you can search more vehicles and catch more insurgents.

Because a vehicle can be carrying hundreds of pounds of explosive it can do damage from a great distance. Including damage to you if you are not careful. See the section on vehicle suicide bombs.

US Military Police search an Afghan local for weapons at a routine checkpoint, near the perimeter of Camp Keating in Afghanistan, September, 2006. (Corbis)

**Mobile Checkpoints:** The purpose of a mobile checkpoint is to capture explosives, weapons and known insurgents in transit as they move from place to place about their business: a temporary checkpoint can be established by either foot or vehicle-borne soldiers but if the soldiers are in vehicles then they can cover a lot more area and, depending on the circumstances, either deny movement to insurgents over a wider ground or catch more of them. The idea is that you are driving along and suddenly stop and form a checkpoint formation – or you leap out of the bushes and begin stopping vehicles.

When you are doing mobile vehicle checkpoints on roads you may turn up all sorts of things and people. A terrorist organization takes a lot of supplying and they have to move weapons and people just like we do. If you think about it with a surprise checkpoint you are much more likely to catch wanted people, explosives and weapons in transit than vehicle bombs primed for use. This is a good thing. Mobile checkpoints are an excellent counter-insurgency tool and can make life very difficult indeed for the insurgent organization if they are done carefully and at random.

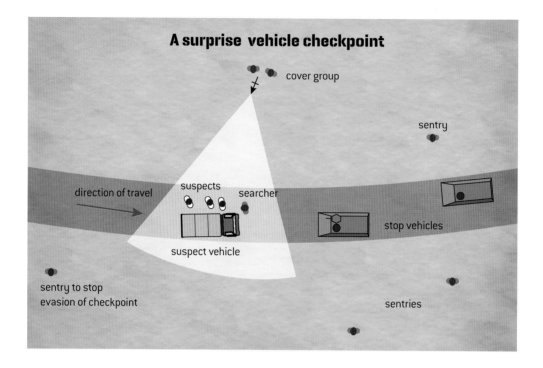

**A surprise vehicle checkpoint**

cover group

sentry

direction of travel

suspects

searcher

suspect vehicle

stop vehicles

sentry to stop
evasion of checkpoint

sentries

I have done countless thousands of mobile checkpoints – many of them with the IRA as the target and I learnt a lot from dealing with them. One lesson was not to underestimate what a member of the public can do when they hate you. On a checkpoint where we had Land Rovers blocking the road one of our men made the mistake of standing in front of the heavy bumpers/fenders. A suspect vehicle came up at moderate speed and ran 'accidentally' into the Land Rover – severing the soldier's legs. The civilian driver claimed it was an accident and got away with it. The soldier was crippled for life. Don't make that mistake yourself.

Often the drivers of insurgent courier vehicles – which are the ones you want to catch – are highly trained. In Ireland I have seen a car do a hand-break turn in its own length on a tiny country lane immediately in front of our mobile checkpoint. Very impressive. It sped away and our rules of engagement did not allow us to fire as we were not at risk so the occupants escaped. If you are not allowed to use fire to stop vehicles a spiked chain works well. Attach the chain to a piece of rope which you lay across the road in such a way that the chain can be dragged out to block vehicles escaping by bursting tyres. Always try to place men well down the road both ways to catch people trying to avoid your checkpoint.

One of the main advantages of mobile vehicle checkpoints is that they are not expected by the courier vehicles. A warning to the driver by signal or mobile phone cancels out this benefit. Because the insurgents know how useful they are to our side, they may have the locals briefed to try to warn the people we are looking for when they see a checkpoint. This means that in an area where the locals support the insurgents you will not only have to move regularly but also watch out for mobs gathering, snipers and sometimes full-blown attacks. Sentries should look 'out' as well as 'in' and beware the efforts of the locals to distract them. On one VCP in the sticks of Northern Ireland a couple of girls did a striptease for us in an upstairs window to take our mind off the job. I suppose it did that.

A few minutes in one place can often be enough to alert the enemy to your presence. The first vehicle you stop may well get on the phone to their local leader and set anything in motion. A bomb for instance. Certainly they will stop the traffic movements you want to catch. Therefore limit your stay to a few minutes at each position. Where the traffic is quiet, start the clock after being seen by a local or stopping your first vehicle. You can sit there all night if no one knows you are there. Mobile checkpoints are a pain for you so make the insurgent suffer too.

## Sentry duty

Most of the time a sentry does nothing at all. But sentry duty is one of the most vital tasks that a soldier can perform. As a sentry you are there to warn the unit which you are guarding of the enemy's actions or approach. The safety of your entire unit is in your hands.

Your unit needs to know as soon as possible that the enemy are approaching, or that they have left a package outside the gate. Whatever the enemy does it is the job of the sentry to make sure the officer in command knows exactly what is happening so he can make the appropriate decision.

Although the purpose is always the same, sentry duty comes in two main flavours defined by the situation and what you are guarding. Basically, you

When the men are tired everyone has to pull a stag, Afghanistan 2010. (Photo courtesy Tom Blakey)

can either be guarding a permanent position such as a building, base or permanent camp or you can be guarding a temporary position – most likely your unit overnight while you are out on patrol. In a sense a sentry is like a checkpoint without the searching because you are denying entry to a secure area. The idea is to let *no one* pass.

There is, or was, a third flavour of sentry duty which I hope you will never come across: where two armies were facing one another along 'lines', as in World War I, then clearly each had to post sentries to watch out for the approach of the other. You will not get this in counter-insurgency warfare as insurgents cannot hold a line against an occupying force. If you do ever find yourself in a classical war then I doubt it will happen then either as modern weapons do not leave much room for infantry holding static lines.

The threat to a permanent position is that the enemy knows where you are and they have all the time in the world to try a mortar attack, a truck bomb, a frontal assault or anything else they can think of.

The threat to a patrol which is just bedding down for the night – be it a foot patrol or motorized – is much more 'off the cuff'. An enemy patrol might stumble across you accidentally, have followed you all day or a local might have guided them in. Whatever the way they find you, you are likely to be facing either a mortar attack – which everyone gets to know about at the same time – or a full-scale assault by a moderately sized force. Of course the idea is that the sentry spots the reconnaissance party before the real show kicks off. This gives you time to make what preparations you can for their reception. The job of the sentry here is essentially early warning.

## Guarding a permanent position or building

We will assume you are in an observation position rather than manning a 'checkpoint'. The difference being that you don't allow anyone in. There are a whole range of places which may be set up for you to do your sentry duty and they all have good and bad points:

■ **Sangar:** Most likely you will find yourself looking out of the observation slit of a 'Sangar' sooner or later. A Sangar is a sort of fortified box made of sandbags or concrete which is provided with viewing slits. The word was first used by the British Army during its time in the North West frontier and Afghanistan in the 19th century – funny how things don't change much. A Sangar's purpose is to protect the occupants while giving them a good field of view from higher than ground level. A well designed Sangar will allow the occupants to look through the slits without crouching or placing unnecessary strain on their backs. If it is difficult to look out  then the average soldier will look out less so if you have anything to do with it, make sure the Sangar has comfortable viewing slits.

- **Tower:** You will certainly get a good view because of the height but towers are often badly designed and vulnerable. They should always have a double skin of protection to defeat the RPG rocket. Be aware that a person entering or leaving may be open to sniper fire. See the anti-RPG diagram earlier in designing a defensive position.

- **Roving patrols:** In some specific cases there is a place for a 'roving patrol' – for reasons other than to mess the men around. If you have a camp made up of huts or tents, and visibility is not good into the lanes they make, it can be a good idea to have pairs of men walking around the area on a random beat. This is particularly true if the perimeter is not as secure as it might be.

- **Crying 'Wolf':** A sentry must never be afraid to tell his superiors when he thinks something is happening. It is far better to check and be wrong than to let something by and perhaps have your base taken by surprise. Commanders should *never* place their men in the position of being afraid to call in a false alarm – even by ridicule.

- **Stags:** There is a limit to how long a man can stay alert. A man will stay alert for a much shorter time than this. At a permanent position there should not be a lack of manpower and a 'Duty Guard' can be created of sufficient size to allow the guards to spend 2 hours 'on' and 4 hours 'off'. 1 hour 'on' is even better. Normally the resting 'off' guards form the immediate response unit in case something interesting happens. In any situation an off period needs to be a minimum of 2 hours long to allow the soldier to benefit from a full 90-minute sleep cycle.

- **Watching the watchers:** The Guard Commander or Guard NCO should take care to visit the sentry positions at random times to ensure the sentries are alert and attentive to their duties. My own opinion is that a sleeping sentry should be shot but as this isn't going to occur certainly some severe punishment should be inflicted.

- **Control room:** If you are very fortunate you may have the benefit of a warm, dry, bug-free control room where you can watch the closed-circuit TV cameras set around the perimeter. I think TV cameras are a great idea as one man can watch many screens, all fitted with night-vision, and a comfortable man is an alert man. There are all sorts of clever technical things to make the sentry's life easier in this situation too – and for the British soldier tea on tap. You see something and you call the guard commander – then he sends out a party to take action. A friend of mine, ex 2eme Rep, French Foreign Legion Parachute Regiment, guards a property owned by one of the richest families in the world. The house is in a large city and he sits 2 weeks on 2 weeks off in a control room watching TV screens. Pretty boring job but terrific money.

- **Guard Commander:** At a base camp an NCO will normally be in physical command of the guard and he will have access to the Duty Officer or Guard Commander as appropriate. Different armies and different units work in different ways but typically you would expect there to be a Duty Officer on duty in any sizeable base and he would have responsibility for taking care of any administrative events which occur at weekends or nights. The Guard Commander is normally the junior officer with direct responsibility for ensuring that the base is properly guarded. These could be the same officer in some circumstances.

## Guarding your patrol overnight

When a small patrol beds down for the night it will often hope to be doing so unobserved. But you can never know for sure. It is an SOP to mount a sentry guard overnight just in case.

The problem is that if you have been walking all day the men will be tired and you want them to get as much rest as possible. To achieve this with a small group of, say, eight men it means that to have two men on guard you will all have to do 2 hours on and 6 hours off. Depending how long you are down for one or more pair may have to do two 'stags'.

In these circumstances it may well be better to have one man on duty at a time so each man only does one stag, or a 1 hour stag, and thereby everyone gets more rest. Of course we don't live in a perfect world but everyone starts to lose their edge quickly when deprived of sleep and particularly if the weather is very cold.

Sentry POV (point of view) behind a 40mm GMG as a Merlin brings in supplies. (Photo courtesy Tom Blakey)

A good leader will balance the need for security with the need for his men to get enough sleep. On very small patrols – such as 4 man special ops – I have often had the men hide and all sleep rather than mount a sentry and have them all tired. Small patrols are easier to hide than larger and tend to be made up of specialist soldiers too. All I can say is that you had better be sure no one knows where you are. You can, of course, set up trip flares and Claymores to stand sentry.

**Choosing the sentry position:** In the old days the sentry used to move and sit behind the gun for his stag as the machine gun had a large proportion of the unit's firepower with the men having bolt action rifles. This would allow the sentry to stop a rush-attack. With the everyone now having automatic rifles this is no longer an issue.

Where the patrol is larger and less well hidden it might be thought wise to have the sentries at a distance where they can detect visitors well before the visitors hear the usual sounds of the camp. You might even want to place a sentry on a high rock or up a tree to give them the chance to see visitors in the distance and at least this way they won't fall asleep – twice.

**REMEMBER:**

As a general rule, if you are going to have sentries then try to have a minimum of two men on guard together as they will keep each other awake, maintain each other's morale and be harder for the enemy to silence.

All things considered it is impossible to give hard and fast rules for the siting of sentries and the commander on the ground must use his experience and judgement to strike a balance between guarding the unit properly, hiding the unit, seeing visitors as early as can be achieved and giving the men as much rest as possible.

# DEFENSIVE POSITIONS: HOLD THAT HILL!

To pull together some of the ideas and principles we have been discussing and try to get them to stick in your mind I am going to take you through a kind of composite mission where you have to hold a hill against all visitors and there is a little bit of everything thrown in to show you how it all fits together. In an actual battle or 'contact' you often can't see the enemy, don't have the right weapons for the job and the rules of engagement stop you doing what you need to do to get the bad guys. Also you would only use some of the kit I mention here and most of the time nothing would be happening – but showing you something *that* realistic wouldn't teach you anything useful would it?

Some famous general once said that in warfare both sides make lots of stupid mistakes and the one that makes the least mistakes wins. Patriotism aside, there is a certain amount of truth in that. When you come to do this sort of thing for real, for God's sake plan carefully, check everything at least twice, rehearse your men until they are doing drills in their sleep and you might just get it right when it counts. But if something can go wrong it will.

**REMEMBER:**

Don't forget the '7Ps': Proper Preparation and Planning Prevents Piss Poor Performance.

*The standard format for a mission briefing is GSMEACS – Ground, Situation, Mission, Enemy Forces, Attachments and Detachments, Commands and Signals. When an officer briefs his men he should go through the information in this order to avoid missing anything out and to aid understanding. For example:*

**Ground:** *The ground is mostly level fields with irrigation ditches at intervals*

**Situation:** *We are driving the insurgents north to clear our squadron area*

**Mission:** *Our mission today is to capture and hold the compound at the road junction*

**Enemy Forces:** *We believe the insurgents number around 50*

*Attachments and Detachments: We have a unit of engineers to put a bridge over the river but our mortar section is missing for this operation*

*Commands: The column will cross the river when I give the command* Water Jump

*Signals: The squadron net will be channel 43 and regimental net channel 55*

*I am intentionally not using this format here so there is no need for the armchair generals to write in...*

## The situation

The country you are working in has a friendly (to NATO), democratic government and is undeveloped economically but almost entirely built on strategically important minerals such as platinum, copper and tin. So the mining rights are a big prize for any developed country. The neighbouring state over the river is very unfriendly to the local government and to ours. They harbour and train terrorists seeking the overthrow of the local government and these terrorists are expected to seek to disrupt the up-coming election to further this aim.

The national border between these two countries is marked by a wide and fast flowing river. There is a steel road bridge but the checkpoint at the nearest end was abandoned years ago because of insurgent action. The country over the river is officially neutral as far as the diplomats are concerned but everyone knows that they allow the terrorists to train on their territory and launch raids across the river border. Our side has high altitude observation drones which overfly them but cannot strike for diplomatic reasons.

The countryside around the friendly side of the bridge is scrubland and there isn't a settlement of any size for 50 miles. Not since the town along the road was burnt out. The locals scratch a living from subsistence farming and a few goats but live in constant terror of theft, rape and murder by the insurgents who come over the bridge in their pick-up trucks, run drugs into the country, destroy government infrastructure and make the locals' lives a misery.

## Your mission

Your mission is to deny the use of the bridge to terrorists wishing to enter the country with road vehicles – pick-up trucks mounting machine guns – during the three weeks leading up to the national election. For political reasons the bridge itself cannot be blown up by our side.

If you are unable to deny the road crossing to the terrorists during the election period their access with their armed pick-up trucks may contribute to the downfall of the local democratic government and its replacement by one sympathetic to the aims of our enemies. In other words, what is happening, as is so often the case in foreign wars, is that you are fighting for our country's interests at second remove by protecting our companies' access to the strategic natural resources mined in this country. This access is granted to our companies by the local government in exchange for military support.

You are aware that should the enemy be intending to cross the bridge with forces sufficient to disrupt the election they must destroy your position first. And the neighbouring government might help them.

## Your troops and weapons

You have been given a company of 140 infantry, a battery of three heavy mortars, a section of three heavy machine guns, three 40mm grenade machine guns and a Javelin missile team with which to carry out this mission. Additions to your command are a light observation drone team and a spook from Intelligence 'Light Side',* aerial reconnaissance, to interpret the Intel it gathers and keep you up to date with the political side of this sensitive operation.

## Your strategy

You have decided the best way to accomplish your mission is to dig in within effective firing distance of the bridge and prevent any enemy forces from crossing – be they terrorist militia or the neighbouring country's army. Simple. To do this, in the face of potentially stiff opposition, it is your intention to create a defensible position which controls the bridge crossing and which you can hold against any one who might turn up. From this position you will be able to bring 81mm mortar, Javelin anti-tank missiles, medium (GPMG) and heavy (.50cal Browning) machine guns, 40mm GMG and sniper fire (8.59mm) to bear on anyone or anything which seeks to cross the bridge.

---

\* Intelligence 'Dark Side' refers to the spooks who work in offices back home. They are allegedly called this because they work in rooms with no windows (for obvious reasons). Not because of their life choices. 'Light Side' are the spooks that work out in the fresh air.

A Javelin Anti-Tank Missile. At $20,000 a shot it has a very clever guidance system and the special warhead will defeat any armour or building. (USMC)

## Your position

You've picked a low hillock to dig in and hold. From this ground, the highest in the area, you command not only the bridge and the road from it but also the surrounding countryside. Visibility extends to approximately 7 miles while you can lay accurate mortar and heavy machine gun fire out to 3 miles, Javelin Missiles, GMG and GPMG fire to 2km, sniper fire out to 1½ km and rifle fire to 300m. Because the ground is relatively smooth and open your position would be a nightmare to assault by unsupported infantry. To take advantage of the only high ground in the area your position has had to be sited 500m away from the bridge — beyond useful assault-rifle range — but your infantry are principally there to protect the position from mass infantry assault. There is nothing like infantry for holding ground. They are as hard to dislodge as rats in holes.

You've had your people dig two-man foxholes in a large circle and put the stores bunkers, GMG, heavy machine gun and mortar positions within it. The Javelin team are free to move where needed and will double as infantry until, and unless, required. In your store bunkers you have rations and water for 28 days and in the ammunition bunker you have reserve ammunition for all weapons. All this weight of stores and equipment was brought in by a couple of Blackhawk choppers under-slung with the mortars and heavy machine guns when the bunkers were complete.

Large pits have been dug for the mortars so the crews have room to work. A separate bunker has been dug on the north side of the position to give the mortar fire control officer a good view of the bridge. He has been using a laser range-finder and GPS to mark the exact range and bearing of various landmarks around the position to increase the accuracy of all weapons.

The GMGs and heavy machine guns have been sited so as to allow the gunners to see their fall of fire all around the position. This has been achieved by placing them on the highest ground with a raised sandbag emplacement. Due to this preparation work the mortars, GMGs and the heavy machine guns can quickly set out ranges and bearings so as to allow for rapid, accurate fire onto pre-selected targets.

There are no trees inside your position and nothing at all growing for 200 yards around. Not now anyway. Just to make sure no one creeps up too close and to stop any possible massed assault, you've put two rings of razor wire around your position. One is 50 yards out and the other 150 yards out. As a rule barbed or razor wire should always be covered by fire and mined if possible as its purpose and effect is to slow down massed troops approaching the position. There are two gaps in the wire where tracks lead to your position and these places are especially well covered by your medium machine guns. You are not using mines as it would take too long to lay and recover them. Obviously, all the infantry have been allocated arcs of fire which interlock with the next-but-one fire team so there is no gap if one is taken out.

Off to the east is a dry stream bed or gully where dead ground* could give cover to an approaching force so you have put some AP mines and a Claymore in it with an electric command wire leading back to the position.

A track runs past the position, also in dead ground, some 200 yards away. It seems likely that approaching forces might use it to close unseen and form up for the assault so you have sent out a patrol to fix trip wires. These will trigger Claymore Mines and illumination flares. That track is not a comfortable place for a night time stroll any more.

---

* Dead ground is an area where the land dips or falls with the effect of obscuring it from view.

An A10 Thunderbolt 'Warthog' letting loose its nose-mounted 30mm explosive cannon at 70 rounds a second. A very useful tool when used in a ground attack and air support role. (USAF)

## Re-supply

The elections are in three weeks and when they are over you can leave. You don't really expect anyone will have a go at such a strong position — you have come so heavily armed to avoid a fight — so you've told your men to make themselves comfortable. You believe you can rely on a chopper re-supply every week for mail and fresh food to keep morale up but whatever happens you are good for ammo, basics and water for a month.

## Air cover

In the briefing you were told you could count on ground attack air-support being overhead in 20 minutes. These could be the fixed-wing A10 Warthogs with their missiles and coaxial roller cannon or they could send Apache attack helicopters also with cannon and Hellfire missiles. The Warthogs pack the biggest punch but the Apaches are more accurate and can hang around longer. But, as ever, you know their availability will depend on what else is going on in the Divisional Area.

## Settling in

It's late afternoon and the men are tired from the digging. You pass the word to make a brew and have a smoke. It will be dark in an hour and if there is an enemy force out there they may try something at night fall. Not an assault probably but maybe the odd mortar round to keep you awake.

At your word the technical sergeant in charge of the observation drone launches his toy. Just as it begins to climb into the air the motor fails and it crashes to the ground with disastrous results. A Techie goes out and recovers the main bits and a few minutes later the reserve drone takes off successfully and starts to do a sweep of the area looking for anything on foot or otherwise unusual. The signal comes back, 'All clear'. There is no one within a couple of miles of the base on our side of the river. It seems likely that any enemy forces would be keeping to the other side during daylight. Where the tree cover is thicker and the drone shouldn't really cross the border.

Just the same, you give the nod to the sergeant and he sends the drone cruising down the middle of the river, scanning the opposite side. A burst of automatic fire from three machine guns hidden in the tree line and the reserve drone goes down in the river. Was it the enemy or was it the other state's regular forces asserting their sovereignty? Is there any difference? Now you don't have an eye in the sky. Was this a diplomatic incident you will have to answer for or was it the first step in a plan to remove you from your watch over the bridge?

## Last light

The light is starting to fade and you have called, 'Stand to'. All your men are at their fire positions. You hear a steady 'thump – thump – thump' as three mortar shells are fired from somewhere in the middle distance. You have no radar to pinpoint their firing positions so you have to sit and take this one without counter-battery fire. The men are all sitting in the bottom

A shadow observation drone launches.
Drones or UAVs are increasingly changing
the nature of SF missions. (USMC)

of their trenches, having a smoke and yet more of the eternal tea. You catch a whiff of what might be rum in the air — but you let it go. The men need to relax when the mortar rounds are coming in.

A crunch and then a pair of crunches as the mortar rounds land. Probably Chinese 82mm. Only one bomb comes down inside the perimeter of the trenches and you hear the shrapnel whistle over your head. Someone shouts something unprintable and not very witty — probably a new recruit. The old soldiers don't make a fuss about mortars when they are dug in. Someone starts to play a harmonica. One of those sad, haunting songs that remind you of home and which are beloved of soldiers the world over. Most soldiers find home looks a lot better from the bottom of a foxhole some thousands of miles away. A shadow whispers as it passes. Just a corporal checking the sentries are awake. The rest of the night passes quietly.

## First light

There have been sentries on watch with night-vision gear all through the darkness but as the first light gives you a view of the mist over the river you have the men 'Stand to' ready for a dawn attack. More tradition from the Zulu wars than anything else it gives the men a sense of organization and rhythm to the day.

The clearing patrol goes out to check for enemy preparing to attack or the tracks of anyone visiting to take a look in the night. When they return it seems a couple of the opposition have come up as far as the 50 yard wire without being seen. The only trace they have left is drag marks as they have slowly wriggled over the soft ground amongst the weeds. Did they come to count you or to mark your position for artillery? The smell of bacon drifts across the camp and mess tins rattle. The men are in good spirits. Life is easy and the weather is warm and dry. Camping is great fun in good weather.

## Routine

The radio operator has picked up a signal from Division to HQ asking why an armed drone was sent over a friendly state's territory. Apparently it had fired rockets at a village full of civilians and killed 17 people. Women and children of course. You wonder if they are talking about your harmless little drone. The argument is raging 5,000 miles away in an embassy somewhere between men in suits drinking cups of civilized coffee. Reality is what is agreed and is therefore negotiable.

A familiar gentle whistling has just become noticeable. It is growing in volume and you know what it is if you can just place it... You scream, 'Take cover!'. The volume goes off the scale as the whistle turns into something like an express train. Then the sky falls, the earth shakes and part of your foxhole caves in.

As consciousness returns you remember what the sound was – heavy artillery. Three 155mm rounds were dropped on your position to make a diplomatic point. Thank God they were just high explosive and not bunker-busters, air burst or phosphorous. You take a walk around the camp and discover there were no casualties but the men look shaken. A reasonable reaction. You report this back to HQ and discover your boss already knew. The opposition were trying to use the made-up drone incident as a cover to take you out entirely with heavy artillery. Your boss heard via a high altitude drone and told them their artillery would be destroyed if they continued firing so it seems there will be no more artillery from the other side of the bridge – probably. Nice to know someone is watching over you and there are some rules to the game.

You've watched the bridge for six days now and there has not been above a dozen mortar rounds fired or a sighting of the enemy or a track that can't be identified. Nothing. You've let the men play football on the killing ground during the middle of the day.

A search team accompanied by some 'Spooks' check out a downed Blackhawk helicopter. (Corbis)

## Re-supply

For all the difference it makes, it's Saturday. Signals report the re-supply chopper is approaching. It's going to be a winch-drop of an under-slung load: water, canned rations, fresh meat and fruit. There will probably be mail too but no more booze than the grunts can smuggle in because the CO back at base doesn't like it. Someone shouts they can see the chopper coming in. It is losing height and approaching your position about a mile away. You call 'Stand to' in case of attack while your men are watching the chopper. The men go to their places cheerful and expectant — every soldier loves a letter.

Three trails of smoke from the river tree line converge on the chopper and it turns into a ball of flame. The ball plummets to the ground a quarter mile from your position. The camp goes quiet. Soldiering is always the same: 99% boredom and 1% terror. You call in the event on the radio and a fleet of gunships come to circle the area while a team search for survivors and classified material. There are none of either. Another chopper, this time covered by a pair of gunships circling high, delivers your supplies before dark. Someone on the other side of the river has decided to raise the stakes.

## Working nights

Another week has almost passed and it has seemed like a month. Nothing but a few stray mortars every night to keep you awake. A week in a place like this is a good rest but much longer and it gets boring. Things could be much worse — a long foot patrol in the sun over soft sand for a start. But you don't want to get too relaxed as a smart enemy always hits you at the beginning or end of a tour. The beginning before you find your feet and get the men trained, the end because you are getting sloppy. Even now night visibility aids are in common use, most fighting still goes on at night. There is still an advantage in the cover of darkness.

A forced whisper brings you scrambling across the camp. The sentry watching the bridge has a sighting report. It looks like there is a gathering of people at the far end of the bridge. Years ago you would have acted on your initiative but now communications have come so far, someone else can carry the buck for this one. Are the enemy massing to come over the bridge or are a group of foreign nationals out for a night-time stroll? Possibly worse, has someone gathered together a bunch of refugees to drive them over the bridge. Are they a human shield?

A video link sends back the night-scope pictures to HQ. As has been traditional for hundreds of years in military circles, HQ comes back with clear, definite orders: 'If it's the enemy you must stop them, if there are civilians amongst them they must not be harmed.' So you know where you stand. Your choice is simple – you either do nothing and risk letting the enemy across the bridge with the consequent failure of your mission or you open fire and risk killing civilians. Ethics is not an integral part of military training. Rather, pragmatism is raised to an art form.

You have the machine gunners loose off a few bursts in front of the crowd to let them know you have the bridge covered. No one is hit but the crowd breaks and you can hear screaming and shouting – then single shots. Looks like the opposition were herding civilians across the bridge in front of them and have shot a couple to calm them down. A radio message comes in that both Divisional airfields are under sustained attack and aircraft are grounded for the time being. It's started.

## Uninvited guests

From all around the perimeter of your camp signals are coming that the enemy is approaching on foot and in some strength. Perhaps 500 men are surrounding you. You give the order and the men hold their fire. It seems the incident on the bridge was a distraction from the approach of a strong infantry force. How did they get within a quarter mile without your sentries seeing them? They knew where the dead ground made a hidden approach possible and it explains why they eliminated the drone. Mortar bombs start to come down on the position in ones and twos. Maybe the visitors the other night were there to get the range right. Certainly their mortar

## TOP TIP!

### For creating and defending your position...

- If you're staying for a while dig a nice comfortable hole
- If there is a chance of armour make sure you have something to stop it
- Use mortars against infantry in the open
- Use GMGs or .50cal Brownings against enemy protected by walls or in buildings
- Sight-in heavy weapons onto likely targets before the off
- Razor wire slows the enemy's advance over the killing ground
- Cover wire with guns and mines
- Stay cool, go steady with the ammo but keep shooting

Mortars lay down defensive fire. (Corbis)

shooting is good and getting better. Enough to make you want to keep your head down anyway and that is just what the enemy wants.

Your mortar commander calls out there is a column of tanks coming over the bridge. They look like Russian T55s – the last-but-one generation main battle tank of the Soviet Union now sold all over the Third World. Not high-tech but reliable with thick armour and a heavy gun. So the nationals over the river are involved. Surely they can't be hoping to keep armour on this side of the bridge? Your aircraft would destroy it in moments – if they were around. But they aren't are they? They are sitting in their bunkers and hardened shelters 200 miles away while the airfield protection teams chase off the guys with the missile launchers.

You decide the opposition are aiming to assault and overrun your position with armoured support and then the armour will return to safety over the river leaving the visiting terrorists with a free hand. And you waiting for the body bags to take you home. There is no air support because NATO Air Forces are under unusually heavy attack at their base. The opposition has brought out some decent missiles – probably black market Stingers – for the occasion and if they are sacrificing these to keep friendly air assets on the ground just to attack your position they must be serious. And you are on your own.

## Assessment

It looks like the opposition are going to make a serious attempt to take your position so they can get men into the country to disrupt the elections. They have tried artillery and, thank God, been stopped by a simple threat of destruction by the high-ups on your side. Now they are pressing forward with an infantry attack supported by ex-Russian tanks. Where the heavy artillery could have destroyed you in your bunkers with a couple of salvos the tanks would take some time and are really to keep your heads well down while the infantry close with you. But they can do that very effectively if you can't neutralize them. You realize that this attack depends entirely on the tanks. If you don't stop them you will be overrun and already the attacking infantry are in their assault positions.

## Contact

Fortunately, they don't seem to realize you have the Javelins. The .50cal Brownings could give armoured cars a headache but hardly scratch a T55. The Javelin, though, was designed to destroy a modern main battle tank and against these old jobs it will be like using a sledge hammer on a peanut.

The tanks are coming over the bridge now – if they fired from the other side there would be hell to pay with Division probably allowing our fly boys to launch a devastating air attack over the river against anything they could find. So the tanks are going to cross the river, take up a position a few hundred yards away and shell you steadily while the infantry advance from the flank. Classic stuff.

The attacking force have set up an 82mm mortar section about a half a mile away and they are starting to lob bombs onto your position. The tanks are over the bridge now and have turned along the road the form up line abreast facing the position. Then everything kicks off at once. You glance at your watch for your Contact Report and it is 21:07.

The Javelin team open up first and take out the lead tank, then the other end to block them in and work their way along the line. As it is hit each tank smokes then erupts in jets of flame and blows its own turret off as the propellant charges and shells inside detonate. In two or three minutes they are all burning wrecks.

As soon as the Javelins open fire so do the GMGs, the .50cal Brownings, your mortars and the rifle sections. The GMGs and Brownings take out the enemy mortars in moments as their heavy rounds cut straight through the sandbag wall protecting the crew.

The mortar crews have the key areas marked on their charts and they are able to bring down a murderous fire onto the attacking forces all around the position. Mortars are no use against men dug in holes but against troops caught in the open like this they are devastating. Stuck between the wire coils the attackers are slaughtered by mortar, GMG and rifle fire.

The few remaining enemy are visible running in the middle distance – until they are brought down by your snipers. You glance again at your watch and the time is 21:21.

It is all over in 14 minutes.

## Summary

Every situation is different, of course, but I've tried here to give you a view of the principle weapons and events that the ordinary infantry soldier might come across in a set-piece firefight within a counter-insurgency situation. This ought to give you a feel of how different weapons are used for different tasks. Maybe not all at once, and maybe you don't have the right ones when you need them, but this little scenario should show you why you have to do some things and how everything fits together. I always think weapons and tactics work like the game of Rock, Paper, Scissors.

Probably the most surprising thing to the soldier new to combat is how quickly things move when they start. Rather than a romantic struggle between two evenly matched teams most battles are a slaughter. This is because one side, normally the one initiating the contact, usually has a decisive advantage in terms of weapons, position, training or all three. An ambush should be over in a few seconds. An air strike on a road convoy the same. About the only time a contact should take a long time is when you are in a strong position and the enemy will not come close enough to be killed but rather take pot shots and loose off machine guns and RPG7s from the middle distance. Even then, you can often bring in air assets to destroy them quickly.

The long part is the waiting for something to happen. For the kick off. I suppose it has always been that way and that the Roman Legionnaires would have felt at home.

## TOP TIP!

## Surviving an interrogation

As a British SAS, US Navy SEAL, Russian Spetznaz or other SF operator you will often be behind enemy lines without heavy support to get you out of trouble. So if something goes wrong you may find yourself a prisoner and subject to an interrogation. In this situation, what can you do to improve your experience? To some extent knowledge dispels fear so let me tell you a little about interrogation techniques. But bear in mind that resistance is only a delaying tactic as a skilled interrogator will always get you to talk in the end.

Most amateur interrogators start with a beating. This is good as it does not hurt much, it may dull your senses for later and it may buy you some time. Remember when Andy McNab was captured by the Iraqis he was held for six weeks and tortured by amateurs. When released he had nerve damage to both hands, a dislocated shoulder and hepatitis – which is not too bad compared to what might have happened.

The **first rule** when being interrogated is to say **nothing** at all during these early stages as a skilled interrogator will start with harmless requests to get you talking then extend these questions into areas of interest to him or her – and in your dazed condition you may not be able to tell what is important.

After a general beating the interrogator may move on to manipulation of your tender spots or the infamous 'water boarding' technique. I experienced a form of water boarding training with a certain foreign unit by having our heads held under water repeatedly. Very unpleasant. Some SF units do teach special breathing techniques to make this more bearable but I'm not going to share this knowledge as a little bit of knowledge can be a dangerous thing in the hands of the wrong people.

At this stage you will start to reveal certain amounts of information. So the **second rule** of interrogation is to start with revealing the information piecemeal (the less important the better) when your pain threshold has been reached but try to ensure this is several hours or days after you were first captured. By then enough time has passed for your own side to be aware of your disappearance and to take steps accordingly to protect any other assets or operations.

And finally, remember that the **third rule** of interrogation is that it is not always about getting information. Sometimes it is simply about making you dread getting caught so it *will be* brutal. Very often you are better dead than captured.

# CHAPTER 10

# Attacking the Enemy

# ATTACK VERSUS DEFENCE

We are finally going to look at how to attack and defeat the enemy — and live to tell the tale. Attacking the enemy is where you will generally destroy them in bulk and achieve any decisive victory. Why? Because you only make the decision to attack when you have some advantage of strength, ground, surprise and/or the enemy is concentrated and available for destruction. Therefore, all things being equal, the attacker always has the advantage and the defender is at a disadvantage.

We have seen a great deal so far about defending yourself in positions, convoys, patrols and so forth when they are attacked. You must always remember that in defence you are really making the best of a bad job because the attacker, in the normal way of things, is only attacking at a time and place of his choosing and this because he believes he can win. Or at least achieve his goal of stopping you resting or killing a couple of men. Far better to take the initiative and attack when you can — the best form of defence is attack.

SF soldiers will often rely on standard forces to supply additional support for large-scale operations. 101st Airborne troops supported Delta personnel during the attack on Uday and Qusay Hussein's compound in July 2003. Both are shown here watching the missile strike on the building which concluded the operation (Delta Force operatives in black helmets). (DoD)

Having said you have the advantage in attack, there are different degrees of advantage depending on the tactical situation:

1. To attack a prepared position it is normal to require three times the defending force because the defenders are dug in and the attacker is crossing open ground. (Air support supplies a lot of attacking firepower and tips the scales overwhelmingly.)
2. When two opposing forces meet unexpectedly then the odds are even unless the ground, training, weapons, numbers or air support gives you an advantage. (You will only prosecute such an attack given you have an advantage.)
3. When ambushing a patrol or convoy the attacker has *all* the cards: surprise, a better position, readiness, everything his way. When laying an ambush greater numbers are not so important because they are replaced by surprise and the preparedness of the attacker which includes the optimum siting of weapons for maximum effect and other factors you will discover shortly. For this reason, ambush is the most efficient form of attack when carried out correctly and should be your preferred way of meeting the enemy.

## But...

To avoid misunderstanding here, let me say something about defending your bases in a counter-insurgency situation: an insurgent attacker ***would*** be at an advantage over you ***if*** he could put together three times the men as an assault team and achieve air superiority ***but he never can***. Therefore he either has to leave your main bases alone or attack with the intention of harassing you as mentioned earlier. In this situation, by efficient defensive strategy, you are effectively making your main bases impregnable by ensuring the attacker can never have a decisive advantage.

# CONCENTRATION OF FIREPOWER

Before we cover the major principles of attack I think it is useful to consider first of all the underlying principle of all combat: concentration of firepower. We will do that briefly below then more thoroughly in *The Assault*. When we have that principle safely in our heads we can

Electronic war games for staff officers. The reality for us guys on the ground is often a little messier. (USMC, Pfc Christopher Lyttle)

look at the idea behind patrolling where, effectively, you are setting out to go into harm's way with more or less the intention of making contact with the enemy.

Once you understand the idea of patrolling, we can look at the most convenient way of destroying the enemy: the ambush – where everything is set up at a time and place of your choosing to ensure you make an efficient job of eliminating all the opposition. Then to stop you feeling over confident about defending any position you happen to hold we will look at how you organize an assault on a position properly. Assaulting a prepared position properly is so important that we will finish off this section with a dramatized demonstration.

In military colleges, historically, there were seven main forms of manoeuvre taught; all associated with the organization of an attack on another enemy force during a set-piece battle like Waterloo. Every one of them was designed to concentrate the fire of part of your army onto a smaller section of the other army and defeat them at that point by achieving a localized superiority of firepower.

Now I'm not training staff officers but what I **am** going to explain here is just what you need to know to stay alive: the principle behind winning every military confrontation you will ever come across. The simple reality is that you win every encounter by achieving a localized superiority of firepower. All ambushes and assaults are based on bringing a localized superiority of firepower to bear upon the enemy. This means that you arrange your forces in such a way that you get more guns in one place pointing at fewer guns to win. Sometimes, of course, this is not rifles against rifles, it is cannon against rifles or air strike against cannon. A bit like rock, paper, scissors. Only different.

## Localized superiority of firepower

What this means in practice is more offensive material going in the direction of the enemy than is coming back. By bringing a concentration of fire to bear on the enemy he can be destroyed, weakened or, at the minimum, his head can be kept down so he stops shooting at you. While the enemy is suppressed in this way the attacking force moves forward to contact without taking too many casualties and eliminates the enemy or captures his position.

Imagine you were part of a dozen riflemen laid in a line shooting at another line of a dozen men just beyond the accurate range of your rifles – say 300–400 yards. If you each fire at your opposite number then you are all going to have about an even number of bullets cracking round your heads and be pretty much evenly matched. The odd man will get hit and each survivor will keep shooting. But if, say, eight of you all start firing at one or two men on the left of the opposite line something different happens – these guys have so much lead coming at them they put their heads down and stop shooting. If you are lucky you might even hit them. This is localized superiority of firepower. OK, in this case nothing much comes of it but the idea can be used to force a breakthrough in the enemy position as you can see below.

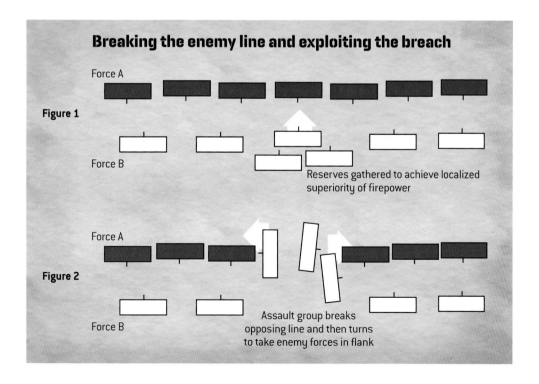

**Breaking the enemy line and exploiting the breach**

Force A

Figure 1

Force B

Reserves gathered to achieve localized superiority of firepower

Force A

Figure 2

Force B

Assault group breaks opposing line and then turns to take enemy forces in flank

The word 'localized' is included because where two large groups are in contact — say two lines of troops facing one another — one group could thicken their line with reserves in one place to achieve a localized superiority there and use that point to break the enemy's line. Once through the enemy line the assault group would turn left or right and 'take the enemy in flank'. This means attack from the side where the enemy are less well defended and are confronted by a stronger assaulting force able to take them piecemeal.

As a general rule of thumb there should be three times the number of men attacking as defending to take a prepared position successfully. This is because it is generally considered much easier to defend than attack as the attackers must often cross open ground to reach the enemy and when there they must overwhelm him quickly to minimize losses. By 'firepower', however, I don't just mean more rifles. If you can bring in gunships or artillery to pin down the enemy while you advance then this has, in theory, the same effect as covering fire from your own machine guns. In practice it is far better.

This is the simplest way I can describe this situation but you will understand the same thing is happening when one side is dug in and an equal-sized force is attacking them — because the defender has the advantage of firepower and cover. The same again in an ambush, the same when an artillery unit delivers a load of HE onto a position and the same when an aircraft drops its load of napalm on an artillery battery. Every situation is the same: the attacker is focusing a lot of ordnance onto a small point, the defender has no reply and is destroyed.

This is why in, say, Afghanistan, your boss flies a few dozen soldiers to attack a small group of insurgents and overwhelm them at that place with a localized superiority of firepower while there are countless thousands of insurgents elsewhere not even engaged in the fight.

Of course, the insurgent leader is trying to gather 200 of his men together without you knowing and then he will attack a little outpost of yours with a dozen of your men in it...

Charging machine guns is not profitable however brave you are: aftermath of a 'Banzai' charge by the Japanese against US Marines, Iwo Jima, 26 March 1945. (Cody)

## Does that mean soldiers are all the same?

However brave, tough, fit and skilful you are personally, the outcome of almost every contact with the enemy is dependant upon who has the drop on the other and thereby achieves a localized superiority of firepower. This is what tactics is all about – getting that drop – and it is the responsibility of the officers and NCOs at their relevant level.

But to bring about a localized superiority of firepower, and even more to exploit one, requires a team of highly trained, disciplined soldiers who will move on command and work as a team. This is your contribution to the victory. That is why soldiers always win out over mobs and gangs of drug dealers etc. Because of the organization and discipline. And when two armies clash, as a rule the best trained soldiers win...

At least they do when they are lead properly. Someone once said that 'War is a series of errors and he who makes the least errors wins'. Someone also said that 'No battle plan survives contact with the enemy'. So when the shit hits the fan, and it **will** happen that you end up outnumbered 20 to one and with no support, that is when you need the tough guys with the guts and the weapons skills. And if they are tough enough they just might hold out till the cavalry arrive.

# THE PATROL

## What is a patrol?

The word 'patrol' is used to describe a force of between four men and generally less than battalion strength (750 men) given a specific operational task. This task might be to seek out and engage the enemy, to blow up a bridge, scout an area or just show the locals you are there and controlling the area. The patrol may last anything from a few minutes for a clearing patrol checking the perimeter of your camp in the morning to some weeks for a reconnaissance patrol. According to the circumstances and operational requirements patrols may be on foot, motorized or placed into area by aircraft.

All patrols are more or less aggressive acts as opposed to defensive. The idea is that you are either setting off to attack the enemy somewhere or going sneakily to do something unpleasant to him like blow up his bridge or similar. For convenience, patrols are usually described by their mission type as this is an obvious guide to their composition.

Patrols are given different names in different armies and some missions can cloud the definition of patrol so bear with me. This list will get the idea across:

- Clearance Patrol is where you check around the camp in the morning
- Fighting Patrol is where you go out to find and engage the enemy
- Reconnaissance Patrol is where you go looking to see what the enemy is up to
- Special Operations Patrol is where you have a job like blowing up a bridge
- Ambush Patrol is where you go out to set up a pre-planned ambush

The preparation for all patrols is essentially the same so we will look at this first then I will cover the differences in turn.

### Necessary knowledge

The basic knowledge and soldiering skills required to perform these patrol functions will have been taught to you in basic training and supplemented by further training in your unit and other sections of this book. The actual techniques, equipment and tactics required will depend to a great extent upon the individual task and the plan of the originating officer.

A very general but useful tip is to make sure that every man in the unit understands the overall plan and aim of the patrol. Then have the men physically practice any significant or unusual actions required to complete the task – such as abseiling down a cliff, crossing a river on floats or the placing of demolition charges. A large proportion of men learn better by 'doing' than 'hearing'.

A soldier's eye view of an infantry patrol – correctly spaced – somewhere in Africa. (Author's Collection)

### Minor tactics (tactics involving infantry only as opposed to an all-arms battle group)

**Spacing:** It is most common to walk in a well spaced single file on patrol. This configuration has several advantages including walking in one set of tracks to reduce the chance of mine injury, lengthening the patrol to make an ambush less effective and ensuring that a mortar bomb, mine blast or whatever will catch the minimum number of men.

Too extended spacing can result in loss of contact and cohesion. The patrol might be difficult to manoeuvre in case of attack. Too close and the problems enumerated above apply. During daylight five yards is normally about right but less may be better to maintain contact in close bush or darkness.

**Contact drill:** The patrol, indeed every soldier, should be well practiced in what to do when the patrol comes under fire. The accepted general strategy is for the group under fire to take cover and return fire while the commander of the patrol arranges for some or all of the remainder to assault the enemy usually in a flanking attack. Of course when the patrol is small, such as a light reconnaissance patrol, it may be the SOP to cut and run in case of contact with the enemy.

When a large patrol comes under fire from a much stronger enemy force the commander may take the decision to disengage by a staggered withdrawal. Rather than forming an assault group the units not engaged may form another fire point further away from the enemy and the group first engaged would then fall back behind them. The withdrawal would continue by repeating this manoeuvre. Where air support is available the commander might choose to set up an instant defensive position and call in air support.

**Separation drill:** After a sudden contact, especially with a poor outcome, your patrol may be split and small groups or individuals find themselves separated. Before the patrol set off the commander should have explained what to do in this situation. During the briefing a patrol's course should be split into what are called 'tactical bounds' and these may be marked as rest stops. Where the exact route is unknown these bounds may be set on the ground. Where a patrol is split it is a SOP in some armies for individuals to make their way back to the last stop.

In other circumstances it may be more useful for the separated members to make their own way back to base or seek an airlift. With the improvement in communications technology leading to many or all soldiers having some form of wireless communication separation has become less of a problem in recent years.

**Tracking:** I believe that all soldiers should be taught the basics of tracking owing to the immense value this skill has when operating in a rural environment. And there is a strong crossover between the skills of tracking and those required for spotting IEDs and mines. There is something of a mystique to tracking and, indeed, experts can look like magicians. I have followed many expert trackers and once followed an African who lead us across scrub

**LEFT** Of course you would be a fool not to hitch a ride when you can. (Author's Collection)
**RIGHT** Around 25 years later and the idea of making the most of local transport is still catching on. Horses and traditional Afghan saddles courtesy of the Northern Alliance, rest of kit and equipment courtesy of Uncle Sam, c.2001. (DoD)

and apparently featureless desert sitting on the front of an armoured car. He could track as quickly as it could drive!

Most of us will never become so highly skilled but careful attention to the ground and surrounding vegetation can, with patience and practice, lead to anyone becoming reasonably proficient in the art of tracking. It is more a case of coming to notice details and getting a 'feel' for the signs of movement over an area. Often traces are noticed subliminally with the intuition leading the tracker forward. In the same vein, a patrol might want to consider avoiding leaving easy tracks when the circumstances warrant the effort.

### Clearing patrol

Each morning it is SOP in many armies to put together a clearing patrol to investigate what has happened around the camp overnight. This patrol might consist of only 2–4 men and they should be looking for tracks or other traces of enemy reconnaissance, they would have the task of disarming 'friendly' booby traps and, after a night contact or firefight with the enemy, they would be doing a body-count and assessing enemy numbers from shell-case piles and losses from bloodstains etc.

### Fighting patrol

This is a group strong enough to take on any enemy forces likely to be encountered and might be given the task of patrolling a certain area and denying it to the enemy. They would be seeking to meet and destroy the enemy by any means available to them. This would include a chance encounter, instant ambush or an attack on a prepared enemy position. The fighting patrol's strength and attachment of specialist forces such as engineers or heavy mortars for an assault would be designed to meet the needs of the situation.

### Reconnaissance patrol

Though replaced to some extent by aerial surveillance, the reconnaissance patrol still has the task of approaching the enemy's position quietly and without detection, discovering strengths and weaknesses and reporting back to HQ. It is usually deployed as a preparation for a ground or air attack. The eyes of a reconnaissance patrol can be aided by the use of a small spotting drone but a reconnaissance patrol can stay in an observation position for weeks where an aircraft cannot.

Coalition SF frequently operate together in Afghanistan. Here US Navy SEALs are shown on operation with New Zealand SAS. The Kiwi is on the left and easily identifiable thanks to his distinctive camouflage pattern. (Photo courtesy of Leigh Neville)

### Observation position

A specific type of reconnaissance patrol is the 'OP'. This patrol fulfils a function sometimes difficult for aerial surveillance – it sits hidden in one place for an extended period to observe movement or whatever is of interest to 'Intelligence'. This might be watching a bridge or river for signs of enemy movement. Normally a small group is used but on a long stint they may work in stags like sentries. On sighting the target the patrol may just report back or they may be required to call in artillery or an air strike. Staying on an OP is, I think, very often an undue risk and a waste of manpower and there are many occasions where it would be more useful for the patrol to set up a video camera with a transmission link back to base and come home. The camera can operate for an extended period, need never sleep and is much harder for the opposition to spot than a bunch of men.

### Special operations patrol

There are various other tasks which may be set for a patrol-type unit. These include the demolition of a bridge or other structure, mounting mobile check-points, escorting supplies or prisoners and supporting the local police in the maintenance of order and public confidence in the rule of law.

Supporting the local police – or replacing them according to the situation – is very much a feature of the modern anti-insurgent function of the military. Where insurgents have compromised the local government or police and temporarily succeeded in their aim of imposing their will on the local populace, it is often necessary to mount foot patrols around the streets not only to catch or deter the terrorist but to reassure the local people that someone is there to protect them and that the rule of law has returned.

## An example patrol

As an example often makes these things clearer I'll tell you about a patrol I took down the banks of the Zambezi river some years ago.

My mission was to ambush and destroy any insurgents crossing the Zambezi river from Zambia into what was then Rhodesia. Zambia was being used as safe ground for training and supply in a similar way to how Pakistan is used today. I had to perform this task for a couple of weeks in different places at my discretion and had just four men, besides myself, to carry it out. Given each group of insurgents might number in the dozens, surprise was of the essence.

The Zambezi is a very wide river and just to complicate things it is thick with crocodile and hippo. I once saw an 18ft crocodile on a sandbank in the Zambezi. And while hippos have a good press, they are actually one of the most fierce and dangerous animals in creation – even the big crocodiles are scared of them. And so are the locals as a hippo will bite a canoe cleanly into two pieces.

The ground cover was moderately dense bush and there were no locals in the patrol area. The ground we were to cover was a huge game park the size of a small country and which had had no visitors for 15 years despite the presence of rhino, elephants and leopards. So the only people who were likely to cross the river were insurgents and poachers coming for the ivory on the elephants and rhino. It was difficult in these circumstances to tell insurgents and poachers apart – but, being fond of animals, I was not inclined to take the trouble.

The routine I set up was to simply mount an ambush on the river each night, shoot anything that came across and then move during the day to a new position. Doing this safely and tactically was a more complex issue.

Each day we would walk down the river for 8 hours or so. Just before dark we would 'dog-leg' back on ourselves to ambush anyone following us and then settle down for a meal – always a mile or so from the river so as not to give away our ambush position. One man would be despatched as a scout to find a good ambush position. The scout would return after dark and lead us into the ambush position he had selected.

The reason for lying up away from the ambush position at last light was to prevent anyone following us into the ambush position and turning the tables on us.

Sundown was at 1830 and then dark followed in 5 minutes. This meant that the scout often had a difficult job to lead the team to the correct place. Sometimes it was a case of

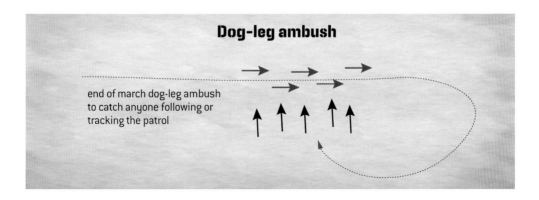

## Dog-leg ambush

end of march dog-leg ambush to catch anyone following or tracking the patrol

holding on to the guy in front. One night we were being led into position in this way amongst bushes when one of the bushes got up and screamed at us. It was an elephant. We turned and ran — straight into another elephant. We turned and ran again into another — we had wandered into the middle of a herd of sleeping elephants. It was so funny — afterwards.

Close by the ambush position four men would set out their light-weight sleeping bags in a cross, heads together, a few yards in from the river bank. This allowed for quiet communication amongst them. The remaining man would be on stag watching the river. Quite a nice job sat in the warm evening — watching the river flow and listening to the creatures making their noises.

A large hippo stumbled on our position one night and managed to hook one of its tusks in my mate Yves Debay's sleeping bag. He shouted in surprise and startled the animal which let out a bellow and set off at a trot. Dragging Yves in his sleeping bag behind it. We almost laughed ourselves sick. Each environment will have its own localized dangers!

One particular night the sentry woke us to say two small boats were coming over the river. We formed up in a line on a cliff some 10ft above the water — four rifles and a machine gun — and riddled the boats from a range of some 15ft in a few seconds. Do you remember what I said about a good ambush being over in 2 seconds? Four rifles and a gun fire over 3,000 rounds per minute — that's 54 rounds per second — and can decimate the enemy.

Then we picked up our kit and ran for a good mile. Mortar fire came over the river onto the site we had ambushed but we were well out of it. The next day we did a recce of the position but all we found was one canoe full of holes washed up on the river bank. The crocodiles had eaten well that night.

# THE AMBUSH

An ambush is using the element of surprise to force a contact on the enemy at a place and time of your choosing and then, hopefully, killing him quickly while minimizing your own casualties.

We have already seen how unpleasant being ambushed can be — even when it is not done properly — so now we are going to learn how to set an ambush efficiently and make it much worse for the enemy. An ambush is particularly useful in counter-insurgency operations when the enemy generally does not want to meet you face to face. It may be the only time you get the chance to kill any of them and do so without the usual required numerical advantage.

The Author and a buddy (photographed left) sat with a radio for three weeks on this hill waiting for someone to ambush. We watched this scene every day for three weeks. And nobody came... Sometimes soldiering really is 99% boredom and 1% sheer terror. (Author's Collection)

The simplest ambush is just waiting, hidden by a track, for the enemy to come along and shooting him when he turns up — but there are quite a number of little tricks I can show you which will make your ambushes a lot more successful and deadly. And deadly for the opposition means safer for you. As a bonus for you, the more you know about setting up ambushes the more chance you will have of spotting the ones set against you. This can only be a good thing.

Every ambush is going to be different according to the terrain, the enemy target and the forces at your disposal to set it up but we will begin by looking at the principles of the perfect ambush.

## The perfect ambush

An ambush consists of a number of component parts. The diagram of a foot ambush below has them all, aside from heavy weapons, while an ambush put together on the spur of the moment may have only a few. If you know and understand the use of all the potential parts you can put together to increase your effectiveness then you can decide, rather like a good chef, when and where to use the ingredients.

In this perfect ambush there is a 'killing ground' behind the enemy so they cannot usefully run away, the ground rises to your position to make it difficult for them to come at you, landmines and Claymores are laid, everyone has been to their positions and can find them in the dark. The main ambush party are waiting, resting, in the Standby Position while a series of sentries watch for the approach of the quarry.

When the enemy are sighted the officer commanding confirms the intention of springing the ambush – as the enemy may be too small or large – then the main party take up their positions.

When the enemy column has moved into the ideal ambush position the commander will give the signal to spring the ambush – usually with a burst of fire. Immediately the front and rear stops fire their Claymores then rake the length of the column with small-arms fire. The main party open up at the same time with their weapons.

Should the enemy column take cover they will either be in the open and easy to hit or trigger the landmines placed in the hollow. If they break and run away from the ambush site the ground has been selected so that they cannot find cover in the open killing ground before they are killed. If the enemy choose to rush the main fire group this should be strong enough to destroy them all.

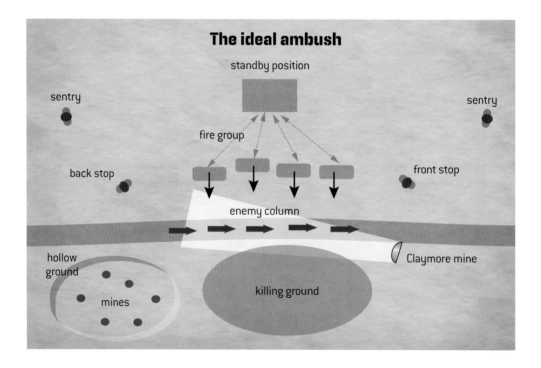

**The ideal ambush**

standby position

sentry

sentry

fire group

back stop

front stop

enemy column

hollow ground

Claymore mine

mines

killing ground

## Terrain

If you are going to sit and wait in an ambush it makes sense to wait where the enemy are likely to pass. You are most likely to catch people walking along tracks or roads but where these are missing from the landscape look for river crossings, passes between mountains and similar features which might be expected to funnel the enemy in front of your sights.

## Killing ground

This should be flat and open wherever possible so as to give a clear view of the enemy as they try to escape.

## Fire group

The position of this should be on higher ground if possible both to give a better fire position and to make an assault difficult for the enemy.

## Front stop

The task of the 'front stop' is to act as a cut-off to prevent the main part of the enemy escaping along their line of march. He should wait off the track in case the commander decides to let people through the ambush site. They may be civilians, or too small or too large a target. When the ambush is activated he should move into position. The front stop may also activate Claymores or other assets with an arc of fire directed back towards the enemy.

## Back stop

The 'back stop' has a similar task to the front stop. He may become front stop if the enemy approaches from the other direction.

## Sentries

Sentries should have arcs of responsibility facing out towards the expected approach of the enemy. They should also be looking out for the approach of civilians and enemy patrols approaching from alternative directions.

## Standby position

The 'standby position' should be used to rest the troops in a long ambush and it should be well guarded in case of attack. Some ambushes may be in place for days or weeks so it should be out of range of camp noise from the track or road. On a long ambush the fire group only needs to move into position when the enemy are known to be approaching. With a large ambush a reserve should be kept at the standby position to deal with attack from an unexpected direction or counter enemy anti-ambush drills with a defensive screen.

## Mines

Where there are hollows or 'dead ground' in the killing ground an enemy will tend to hide there when he comes under fire. It is often a good idea to accommodate the enemy by laying landmines in these places.

It has been known for a weak force to open fire on a stronger force in an apparent ambush with a killing ground providing plenty of cover. The immediate action of the ambushed force has been to take cover only to find that explosive charges and mines have been placed in all the positions providing cover. Effectively, the larger force has been driven onto the IEDs of the smaller force where they may be destroyed at will by command detonation of the charges.

This is what a Claymore looks like when it goes off... (USMC, Lance Cpl. James W. Clark)

Claymore Mines provide a very useful directional blast as has been explained elsewhere. As the best time to hit the enemy in an ambush is in the first couple of seconds, before he has chance to take cover and move out of line, the Claymore is useful for directing blast and shrapnel along the approach path of the enemy when fired by command wire.

> **REMEMBER:**
>
> Whenever you are in a vehicle which is hit by an IED or stopped by a blockage in the road assume it is an ambush and react accordingly.

## Vehicle ambush

A vehicle ambush is in most respects identical to the ambush of a foot patrol. If the vehicles are armoured then clearly heavier weapons are required. One difference is that vehicles often move at speed and therefore the front stop needs to be in a position to stop the column quickly or, failing that, fire along the length of it to good effect. It is also the case that vehicles often cannot escape across country and therefore may be trapped more easily than foot soldiers by mines or IEDs immobilizing the lead and rear vehicles. Stop the front and rear vehicles and you often have them all.

# THE ASSAULT

In counter-insurgency warfare the enemy is, by definition, weaker overall than the occupying power or you wouldn't be doing much occupying. This insurgent weakness is often not in terms of numbers or bravery but results from a shortage of air assets which leads to our air strength creating a shortage of ground assets for them. That is, if the enemy ever had an organized army, our air assets have stuffed his air assets and then cut up his organized ground assets such as troop formations, tanks and artillery leaving him hiding in small groups merely to survive.

This means the insurgent will attack isolated groups of our forces, if he can manage to gather sufficient forces of his own, before melting away into the hills. Alternatively, or as well, he will resort to bombs under the road and other tactics which don't give you much chance to shoot him.

**TOP TIP!**

## Mix and match

You may not be able to bring all the potential assets listed above to bear on every ambush but the more you can use the more successful you will be. A successful ambush is one where the enemy don't get to shoot back at you before they die.

In an anti-insurgency role you will find that the ambush is probably the best way of bringing an enemy to contact who is reluctant to face you in open battle.

The straight forward assault, a direct attack on an enemy position, is one of the most satisfying methods for eliminating the insurgent. Of course, in a counter-insurgency situation, the enemy is not intentionally going to be sitting there waiting for you to attack him so the opportunity to assault his position generally arises in one of two forms which I will label the *Spontaneous Assault* and the *Prepared Assault*.

## The spontaneous assault

This occurs when you survive an ambush set against you or you bump into the enemy while out on patrol – **and the enemy does not run**. You then use your forces to organize an assault, close with the enemy and destroy him.

## The prepared assault

This occurs when our Intel people have found out where there is a bunch of insurgents staying for a period in a house, compound or camp. This may be in a neighbouring country to the one you are occupying as such a place is often used as a safe haven for training and resupply in insurgency campaigns. In the occupied country the insurgents may have a hidden supply depot, training camp or just be taking some R&R in what they think is a safe area. Again, you use your forces to close with the enemy, organize an assault and destroy him; but this time you have more time to think and gather what you need for the job.

First of all, so you understand where the principles come from, we will look at a prepared assault in a classical warfare situation as opposed to an anti-insurgency campaign.

British Parachute Regiment patrol EMOE a compound wall and enter, Afghanistan 2010. EMOE is 'Explosive Method of Entry'. (Photo courtesy Tom Blakey)

## The standard assault technique

In classical warfare, a prepared assault is a planned and organized attack on an enemy who is in a, more or less, prepared position therefore most assaults are against trenches, bunkers or buildings. Obviously, the intention of the enemy is to make these positions difficult or impossible to capture by means of trenches and protected fire positions so a special technique has to be employed in order to take them — or at least without a horrendous loss of life on our side. The last thing you want is to be part of a wave of men sent 'over the top' to be shot down as you approach the enemy position as in World War I. Pay careful attention here so you will know when it is done properly and when some officer just wants to win medals.

I will say again, because it is so important, that to assault a position successfully you need to be able to bring sufficient fire to bear on the enemy to at least keep their heads down. Obviously it is better if you can kill or demoralize them but keeping their heads down allows you to move forward without being shot. This firepower does not all need to come from your own infantry. You can use mortars, artillery, air support or anything else you can bring to the party to suppress the enemy.

For the assault technique itself an example will make all plain: let us suppose you are in company strength (120 men) and assaulting a platoon strength (40 men) enemy position dug in on a hill. I realize that some old soldiers will laugh at my optimism as the British and US armies are often expected to attack without such odds in its favour – like attacking a prepared position with less men than the defenders had in the Falklands – but this is how it is supposed to be done and lack of numbers can be made up by heavier supporting weapons – if you are lucky.

**Step 1:** Split your forces into a fire group and an assault group. The fire group, amongst infantry on foot, are generally medium machine guns (GPMGs) but if you have vehicles they may be .50cal Brownings or Grenade Machine Guns (GMG). The fire group has the task of keeping the enemy's heads down while the assault group – the riflemen – advance.

As you don't want the fire group shooting your assault group when they reach the enemy position or on the way you need to separate them and their arcs of movement and fire. This is achieved by either keeping the fire group in front of the enemy and sending the assault group out on a flank to attack the enemy from the side, this is called a 'Flanking Assault' or the reverse, where you send the fire group out to the flank and have the assault group attack the enemy from the front. The latter is called a 'Frontal Assault'. These are the official terms but in my opinion these two forms of attack are effectively the same – all you are doing is sending your fire group to the best fire position and your assault group to the best assault position given the lay of the land with a 90° of arc between them.

Make your choice of position for your forces according to the ground and the defences the enemy have set up – clearly if you can get your fire group above the enemy on a nearby hill then they can see their targets better. In this example you are performing a frontal assault with your gun group out to right flank as fire group. Your fire group brings fire to bear on the enemy and suppress his fire against your approaching assault group. That is, they place the beaten zone of their weapons on the enemy position and keep up a constant fire.

**Step 2:** As the assault group reaches the enemy position the fire group switch fire away from the position so as to avoid hitting your own men. They continue to fire because the sound of munitions cracking over his head can still suppress the enemy as the assault group

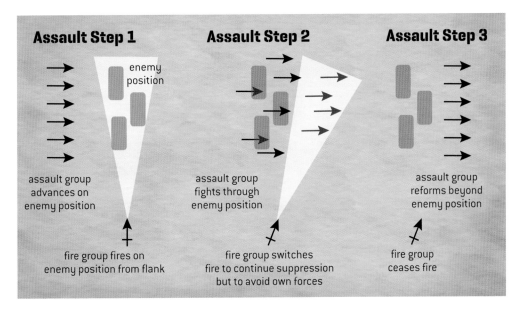

**Assault Step 1**

enemy position

assault group advances on enemy position

fire group fires on enemy position from flank

**Assault Step 2**

assault group fights through enemy position

fire group switches fire to continue suppression but to avoid own forces

**Assault Step 3**

assault group reforms beyond enemy position

fire group ceases fire

makes contact – particularly if the enemy defence is deep and if they have further rows of trenches behind the front line. Then your assault group fights through the position with grenades and small arms.

**Step 3:** The assault group continues on through the enemy position and out the other side to re-form into a defensive fire position. You should always do this in case enemy artillery or air is targeted on the enemy position to spoil your celebrations. The fire group cease fire. When the enemy has failed to shell their old position the assault group send back a party to the position for a body count and to collect ammunition, documents and maps etc.

There are two things I want to bring to your attention here. The first is that the fire group could be a machine gun or it could be a group of mortars, a squadron of bombers or a regiment of artillery depending on the size of the enemy and the tactical/logistic situation.

The second point is that this type of assault, even when set up properly, involves closing with a potentially very lively and aggressive enemy who will kill you if they possibly can. It can lead to heavy casualties on the attacking side.

## Modern thinking on assaulting enemy positions

Since World War II technology has moved forward and armies have become better supplied with more powerful weapons. This means, in practice, that we, the Western Alliance or 'Good Guys', have lots of air support and clever, accurate, long-distance artillery. Don't laugh, I know it never seems to turn up when you need it but when I was a boy the infantry had no air assets to call on and the artillery were always too busy or out of range.

Society has moved forward too and it is no longer thought to be acceptable to send lots of our young men walking towards machine guns until the enemy gunners run out of ammo through shooting our men and are overwhelmed. So tactics have been developed which seek to cut down our casualties by not giving the enemy the chance to shoot us. This can only be a good thing for us soldiers.

As far as assaulting a position is concerned, and given you have access to air or artillery support, the latest best practice is to bring infantry close enough just to pin down the enemy and stop him escaping. This, of course, can generally be achieved without coming close enough to get shot. Then we call in air or artillery and blow the enemy to pieces. Great work when you can get it.

The significant difference here is that rather than splitting your forces and using the fire group to suppress the enemy while the assault group comes to grips, the whole of the infantry team is used as a fire group to pin down the enemy while the assault group consists of more advanced weapons such as attack helicopters, ground-attack aircraft, heavy artillery shells or bunker-penetrating bombs to entirely destroy the enemy. Then all you have to do is go in and count the bodies. So the modern way to do it is:

1. Find the enemy through intelligence, electronic means or boots on the ground
2. Stop the enemy moving by pinning him down with ground troops
3. Destroy the enemy with artillery or air power but no assault troops
4. Send forward ground troops to search the position

## Clearing a position

When you advance on a position which has been suppressed by fire of whatever kind your main goal is to kill anything in that position before it can kill you. There may be people in trenches with their heads down and weapons in their hands waiting to kill you. There may be people cowering in the bottom of trenches or bunkers, there may be wounded and there may be people trying to surrender.

Unless weapons have **clearly** been discarded and hands are in the air, kill them all as you advance line-abreast through the position unless you have specific orders to obtain prisoners for questioning. You cannot tell on the spot if a person is genuine or not, armed or not and you have to wonder if they really have your best interests at heart. If you stop to try to take a prisoner or whatever you are breaking the line of your men advancing through the position and thereby putting everyone at risk. Move through the position using rifles and grenades and leave no one alive. Regroup at the other side in a defensive position and send a team back to search the bodies.

## Clearing a building

At times you will have to assault or clear buildings. This is particularly difficult to achieve safely. You will doubtless have seen heroic Special Forces soldiers or SWAT police on TV running into buildings and rooms with torches strapped to their weapons and fancy masks on their heads. This tactic is either for TV or to give the enemy a fair chance at killing you. Avoid this at all costs. The way to clear a room is to toss in a grenade before you go in. Either a hand grenade or a rifle launched grenade is fine. And then, if you are not sure, toss in another.

You are only clearing this room or building because there may be enemy in there and you cannot destroy the building with explosives. Otherwise why would you want to go in? If the room is empty then no problem. If there is a non-insurgent inside then they really ought to come out with their hands up or their day will get worse.

Your problem arises when there is someone in that room who has evil intent towards you. However heroic, well armed and well trained you are, all they have to do is sit still and shoot you through the door as you open it or as you walk in looking around for them. They know exactly where you are – the doorway – but you don't know where they are. Especially if it is dark and you have a torch in your hand or on your rifle. You are at a massive disadvantage and will die when you meet opposition. Even if you have heavy armour, it doesn't cover all your head and body or stop every weapon so how lucky are you, really?

## Pointers for staying alive in urban warfare

Given you cannot always use an air strike or artillery to flatten the place here are some tips:

To clear a whole building start at the top, if you can, and work down so that you can drop grenades down the stair-well ahead of you if required.

To clear a room, toss in a grenade through the window if you have to approach from ground level. If not, shoot up or down through the floorboards before throwing in a grenade. If it seems sensible, have someone outside with a grenade launcher lobbing in the grenades through the window while another team, with careful timing, clears the rooms. If you open the door to toss in a grenade someone will shoot you through the door.

If you have the knack, time your grenades to explode in the air for best effect. The flash, bang and shrapnel from the grenade will stun anyone in the room long enough for you to walk in and finish them.

On entering a room, if you are using a rifle aim for the centre of the body for a sure kill. If you are using a pistol or SMG aim for the head if they are not moving otherwise the body first then the head as they may be wearing body armour as underwear.

If you are fortunate, you may be issued with some handy little explosive charges called 'door knockers' by the British on the basis that they get doors open. Sometimes doors are remarkably difficult to kick in. And even when they are not, would you really want to be kicking a door with an armed enemy on the other side? If it were me I would shoot you through the door. Place one of these charges against the door and set them off then follow up with a grenade to make sure. The door will blow inwards and the noise will help pacify the occupants. The US Army have a whole series of useful toys from special rifle grenades and heavy duty shotguns to tailor-made explosive charges. Don't you just love American kit?

> **REMEMBER:**
> If you run into a room with a gun, the person already inside has the advantage and, if they are not stunned or timid, they will kill you before you get a bead on them.

If there are 'innocent' civilians inside the chances are that a grenade will do them no permanent harm but when you enter do be careful not to leave a bad person alive just because they are female — there are plenty of armed and dangerous female insurgents.

## Tactics of the future

As technology develops we are going to be issued more advanced robotic weapons. Tactics will be modified to keep pace and hopefully keep us alive longer. Already we are starting to see a situation where infantry work as spotters to call in fire from air and artillery when they find the enemy.

And we all know about the armed drone aircraft such as the Predator and the Reaper which are flown from bases in the States and are being used to target the bad guys behind enemy

This is how to open a door – there is no need to be polite. (Corbis)

lines without risking our pilots. And robots which creep up on bombs and disarm them have been around for decades.

Within a few years we will be supplied with their grandchildren: there are already 'drone' armoured vehicles in testing. Effectively these are small tanks armed with roller cannon which can either be controlled from base or can 'think' for themselves. Does anyone else find that a bit spooky? I have seen these and they look like something from a science fiction movie. The thing is though, they kill the enemy without exposing you to fire. That can only be a good thing.

I can see tactics evolving where there are actual patrols of these 'ground drones' working across the countryside in Afghanistan and elsewhere. Back at base, in the States or wherever, each vehicle would be operated by a commander, a driver and several gunners all sitting at computer screens in a shed and playing what look like computer games. Very much like what the air drone pilots do now. At this point we are un-killable as there are no men involved on our side yet the land drones can do everything a vehicle patrol can. They could, theoretically, even speak with the locals over a video link. I am developing the tactics for use with ground drones.

I have seen a robot sentry gun turret manufactured by a South Korean firm which is a step on from the radar-controlled roller cannon used currently to shoot down rockets and shells approaching ships and our larger land bases. This robot gun turret mounts a .50cal Browning – though it could be a GMG or any cannon – and watches its target arc for enemies using light, infra-red and other sensors to detect humans or moving enemy assets. Then, when it sees something it doesn't like the look of, it opens fire without a human operator. And then it corrects its fire until it hits the target. I saw one of these destroy high-speed boats approaching the shore and targets on land. This is the future.

# TACTICS FOR THE ATTACK: DESTROY THAT ENEMY BASE!

We have now looked at the principle ways of attacking the enemy and I hope I have made the important points clear to you so that you understand what is required and can either put a reasonable plan of attack together; or at least understand and follow someone else's plan intelligently.

It is my belief that if everyone involved in an operation understands what the objective is and how it is going to be achieved then this makes all ranks more motivated and, given how often things go wrong in the face of the enemy, more able to find other ways of achieving the same overall objective. Of course, you can only tell your men what is happening so far as is possible given security and other constraints. You wouldn't want someone who may be captured knowing your long-term plans would you?

As I have said before, some people learn best by being told how things

Don't be a hero, call in the big guns of aerial support when you can: gunner's camera on an Apache attack helicopter. Lurking high in the dark sky and some kilometres away it can easily destroy individuals on the ground. (Corbis)

work, some learn by remembering all the facts and some people learn best by seeing a working example. To try to cover all our option here I shall finish off with a little story about a fictional attack on an enemy position. It will seem a little 'busy' but I have tried to involve as many of the options and possibilities as I could so you will see where they all fit in.

*The standard format for a mission briefing, as I have shown previously, is GSMEACS – Ground, Situation, Mission, Enemy forces, Attachments and Detachments, Commands and Signals. Don't think I have forgotten, but some of this standard information is irrelevant here as this story is more than a briefing and continues through the actual contact to the end of the mission.*

## The situation

The Western Alliance is occupying a country similar to Afghanistan on the pretext that we are bringing freedom, democracy, education and the emancipation of women to the locals while preventing Islamic insurgents and extremists from training and organizing attacks on Europe and the USA. We will refer to the country we are occupying as 'Country A'.

While there is some definite truth in this we also intend to control an oil supply pipeline which runs westwards across the north of landlocked Country A to a West-friendly sea port in a neighbouring country and also gain access to strategic reserves within Country A which include 36 trillion cubic feet ($1,000 \text{ km}^3$) of natural gas, 3.6 billion barrels of petroleum and up to 1,325 million barrels of other natural gas liquids. Recent reports also show huge amounts of gold, copper, coal, iron ore and other minerals.

A few years ago we crushed the Third World defences of Country A in a conventional blitzkrieg. First our air assets destroyed their conventional assets and then we occupied the country. After which an interim government was installed. Almost immediately after the new government took over a strong Islamic insurgency movement sprang up and made Country A impossible for the local government to control without a huge and costly Western military presence. The aim of the insurgent foot soldiers is to throw the Western invading infidels out of their land and install an Islamic government while the leaders of the insurgency want to control the oil pipeline and have access to their country's mineral wealth.

Since taking over, the Western Alliance has built up an occupying army of over 100,000 men, not including civilian military contractors, and is making life difficult for the insurgents by active patrolling and air strikes against targets identified by agents, satellites and drones. Our unquestioned air superiority means that the insurgents are unable to field a ground force of significant size without our air assets destroying it so they are laying mines and IEDs against our foot patrols and supply convoys, sending a steady stream of suicide bombers to kill civilians in markets or soldiers at vehicle check points and destroy any government buildings they can gain access to. There has in recent weeks been a series of effective attacks against our helicopters using Stinger type ground-to-air missiles and this is causing serious problems in both the supply and the support of our ground troops from the air.

IEDs killing our men on the ground and effective missile attacks against our helicopters are widely reported on the international news and this is causing difficulty at home by raising a groundswell of voter opposition to the occupation. If the insurgents' attacks are not rendered less effective they may lead to a total withdrawal of our troops and the loss of resources vital to the West because, without our intervention, these would be annexed by China or Russia.

To the south of Country A there is a country similar to Pakistan. We will refer to it as Country P. This country is playing a double game: they are friendly to the West on the surface – as otherwise we would invade or destroy them on one pretext or another – but there is a strong sympathy for the insurgents and a wish to aid them in driving out the invader. But of course they must be careful not to be shown to be aiding the insurgents. If the West could be removed from Country A then Country P would be able to exploit the natural resources of the country by supplying access to deep water ports for the export of minerals from Country A, a major advantage to a relatively poor country and also a way to increase its own power in the region.

A standard tactic of insurgents the world over is to use camps in a neighbouring country for training, resupply and R&R. Insurgent missions are launched from this neighbouring country and afterwards insurgents escape there if they can. Usually the occupying forces cannot destroy bases in a sovereign foreign country nor chase insurgents into their territory.

In this particular case the government of Country P claims to have little control over those areas of its own country which border Country A but refuses to give Western Forces access. These wild and mountainous areas are being used by the insurgents of Country A for all the purposes listed above plus these same insurgents are carrying out attacks on our supply convoys passing through Country P on their way to Country A. While they are attacking our supply lines almost with impunity, the military of Country P are doing little to stop them.

For some time now our generals have been sending unmanned drone aircraft such as the Predator and Reaper attack drones to find, observe and destroy insurgent camps and safe houses inside Country P. These have been supported by local agents on the ground but, as

## Reaper spec

Drones are changing the way SF troops fight battles. The Reaper is one drone that is leading the way and you will become increasingly familiar with. Target acquisition includes the Raytheon AN/AAS-52 multi-spectral targeting sensor suite, a colour/monochrome daylight TV, infra-red and image-intensified TV with laser range-finder/target designator which can designate targets for laser guided munitions. The Synthetic Aperture Radar system enables GBU-38 Joint Direct Attack Munitions (JDAM) targeting, is capable of very fine resolution in both spotlight and strip modes and has the capability to indicate ground moving targets. It can carry up to 14 AGM-114 Hellfire air-to-ground missiles or four Hellfire missiles and two 500lb (230kg) GBU-12 Paveway II laser-guided bombs. The 500lb (230kg) GBU-38 JDAM are another optional extra. Testing is currently underway to support the operation of the AIM-92 Stinger air-to-air missile for use against unfriendly air assets.

Reaper MQ-9 drone carrying about four times the bomb load of a Predator. (USAF)

ever, the targets the agents supply are sometimes good but sometimes chosen to settle local scores; and occasionally they are selected to destroy a compound hosting a wedding or similar to strengthen anti-Western feeling.

Our Special Forces have, on occasion, been sent into Country P both in hot pursuit of terrorists and in small raids just over the border. The limiting danger has always been that a diplomatic incident would be caused by the capture of any Western soldier inside Country P, and Country P is, quite understandably, already angry about our incursions.

A report has just come in that there is an insurgent base 50km inside Country P which is both a training camp for insurgents and a storage facility for the Stinger anti-aircraft missiles which are being supplied to the insurgents by Country P.

Our generals consider it vital that these missiles are destroyed to ensure the safety of our helicopters. In addition if proof can be obtained that clearly shows that Country P has supplied the missiles then our diplomats can prevent the supply of further missiles. Drones or bombers can destroy a camp from the air but they cannot collect paperwork or take prisoners. Only ground troops can do that.

## Your mission

Your mission is to lead a team of Special Forces into Country P, destroy the insurgent base and obtain proof that Country P is supplying these anti-aircraft missiles. The survival of our entire helicopter force depends on your success as does the safety of the 100,000 men they support with transport, supplies and casevac. No pressure there then.

## Your troops and weapons

The force you use will comprise a mixture of British and Australian SAS, US Navy SEALS and other SF plus US Intelligence and language experts.

Fixed-wing ground attack and rotor craft are available to you plus any special weapons or munitions you may require for the task.

## Your strategy

Your plan is quite simple: you will split your forces into six eight-man fire teams, Alpha 1–6, plus the Intel and language people, Zulu 7, which is four men plus four bodyguards. Each team will be drawn from only one unit so they are used to operating together.

All your teams will HAHO* in from a 25 mile standoff to the target and take up positions to block roads into the target while assault groups clear and search the base.

Once on the ground, your forces will clear missile stores and command offices of the enemy then destroy these after photographing the area and searching for paperwork on the missiles.

Reaper drones will continue to patrol the area of the base throughout the operation to prevent reinforcements and be on call to the stop groups.

Extraction will be by a squadron of Blackhawks when the operation is over and the area is secure.

You aren't the only ones who will come to the party well armed. Insurgents with Stinger ground-to-air missiles. (Corbis)

## Execution

It is two hours before dawn and the inside of the C-130J Super Hercules is loud. For the last hour before jumping the jumpmaster signals for all the men to switch to breathing pure oxygen. This is to reduce the risk of the 'bends' caused by nitrogen bubbles forming in the blood on decompression – a particular problem with HAHO jumping when a long traverse is made. You check that every man has switched.

Just over 30 miles from the target the red light comes on in the Hercules and you stand up. All the teams check each other's equipment. They move to the exit door and watch the jumpmaster. The green light comes on and everyone jumps in teams of eight with a brief gap between each team. You jump last.

A few seconds out of the aircraft and the square canopies open in groups of eight. A little circling and you all set off flying further into Country P, still over radar. Each team leader navigating by the GPS on his chest mount.

---

\* HAHO stands for High Altitude High Opening and refers to a means of parachute insertion where the operators jump at high altitude over the enemy radar and then open their square canopies while still at a high, and classified, altitude allowing them to fly undetected for an extended distance in excess of 30 miles to their target, all the while navigating by chest-mounted GPS.

Preparing to parachute for a night-time HAHO insertion. (DoD)

At 6,000ft you see the teams on your flank begin to veer away as each heads for its allotted position. Seconds later you begin to circle a few yards from the target and your team lands together with the Intel crew. Dawn is lighting the sky across the horizon. The other side of the Drop Zone, but out of your sight, the remaining teams touch down almost together and the men set about redistributing the equipment they are carrying back to what each man needs for the operation. Mainly ammunition and explosives. (On HAHO operations it is vital that each unit of man-plus-equipment weighs the same so that all operators in a team descend at the same rate.)

You press to speak on the radio, 'All stations Alpha and Zulu this is Niner radio check over'. The replies come back smartly, 'Alpha 1 Fives over, Alpha 2 Fives over...' All the teams can hear you just fine. A nod and you lead Alpha 9 and the Intel crew of Zulu 7 towards the office block, a wooden shed some 40ft long. There are two guards sitting dozing, their backs against the wall. You all squat down against a low wall.

You use the radio again, 'All stations Alpha confirm position over'. The replies tell you they are in position. Alpha 1 is watching the west road with Alpha 2 in reserve, Alpha 3 is watching the north road with Alpha 4 in reserve. Alpha 5 is in position to enter the missile hangar and Alpha 6 is ready to support them. 'All stations Alpha stand by.'

You take a last look around and all seems quiet. 'Alpha 5 move now, over.' 'Alpha 5 moving now, out.' Alpha 5 are entering the missile hangar.

A nod to your sergeant and two shots from a silenced 9mm Glock SMG kill the guards by the office. Time to call in the Reapers. You switch frequency and call the Reaper controllers some thousands of miles away back in the USA. You are able to do this because your signal is relayed by one of the Blackhawks above. 'Reaper 9er this is Alpha 9er fire now over.' 'Alpha 9er this is Reaper 9er firing now.' Almost together 4 flaming trails mark the flight of Hellfire air-to-ground missiles from the Reaper Drones high above you down onto the accommodation blocks of the base. Each is totally destroyed by a deafening blast.

'Ach, that's the way you motherf**ker.' A Glaswegian Scottish accent expresses satisfaction at the destruction of all the enemy troops without their firing a shot. 'Shut your face' – your sergeant has the interpersonal skills of a wolverine. You nod to the sergeant, he points to the Glaswegian and another man. They jog to the door of the office, open the door quietly and slip inside. Your sergeant and the remaining team members take up positions on either side of the building. An individual climbs out of a window and is shot silently from outside. Lights go on inside and you turn to the Intelligence officer, 'All yours Jeff.' The Intel captain leads his team into the building to look for incriminating paperwork or anything which might tie the missiles to Country P.

Suddenly a number of automatic weapons open fire from the direction of the missile hangar. 'What the...' You turn to look and can see bursts of fire inside the dark hangar. Your radio bursts into life, 'Alpha 9er, Alpha 5, contact at the missile shed, looks like 30 or 40 men were sleeping there and we have woken them up. One man down, over.'

You reply quickly: 'Stations Alpha 5 Alpha 6, this is Alpha 9er. Withdraw from the hangar and see if the enemy follow. Alpha 6 ambush them if they do, over.' 'Alpha 5 copy, out.' 'Alpha 6 copy, out.'

Jeff calls you: 'Alpha 9er this is Zulu 9er over.' 'Alpha 9er send over.' 'Zulu 9er we have all the papers and photos we need from here. Am leaving the building.'

'Alpha 9er copy, return my location out.' You order the Intel crew to come back to your position by the office building then tell Alpha 5 and Alpha 6 they can head for the extraction site. 'Alpha 5, Alpha 6, abort ambush at your discretion and RV extraction over.'

'Alpha 5 copy, out.' 'Alpha 6 copy, out.'

Now you have the papers the missiles can be destroyed... 'Reaper 9er this is Alpha 9er target missile shed, over.' 'Reaper 9er confirm target missile shed, over.' 'Alpha 9er confirm target missile shed, over.' 'Reaper 9er copy missile shed, out.' A few seconds later two lances of fire streak down from the sky and the missile hangar erupts in terrific sheet of flame.

## Extraction

'All stations Alpha RV extraction, over.' The Alpha callsigns reply that they have heard and will comply. 'OK guys, let's head for the extraction point.' Time to call the taxis, 'Heelo 9er this is

## SAS Operation *Marlborough*

Operation *Marlborough* was not as glamorous or newsworthy as some SF operations but it was planned and executed with such precision and elegance that it is certainly worthy of recording here as an example of how to do things properly.

In July 2005 British Agents working for MI6 discovered a plot in Baghdad, Iraq for several suicide bombers to detonate their explosive vests at various restaurants and cafes frequented by the Iraqi military. The terrorist base was quickly identified as a certain house near the targets and listening devices were placed in the walls to overhear conversations inside.

It was soon determined there were to be three suicide bombers cooperating in a synchronized attack. In addition, within their house was a bomb factory and an unknown amount of explosives.

Task Force Black, a joint US/British Special Forces unit comprising SAS, SBS (Special Boat Service) and Delta Force, was given the job of eliminating the terrorists and a plan was devised where the British SAS would kill the terrorists while British and US troops provided support and cut-off groups in case of unforeseen events.

To kill the terrorists four two-man SAS sniper teams would be deployed on rooftops surrounding the house and take down the bombers as they left the building. That is three

Alpha 9er, extraction minutes 10, over.' The Blackhawk commander comes back: 'Heelo 9er Confirm extraction minutes 10 out.' You set off at a brisk walk followed by your team and Zulu callsign.

Seconds later a call comes in over the radio, 'Alpha 9er this is Alpha 1, a convoy is approaching our location from the west at speed. Distance 3½ miles. 12 vehicles, company of infantry and light armour, over.' You reply, 'Alpha 9er, copy that. RV extraction out.' Looks like someone is coming to take a look. Probably the locals will not get involved but you had better warn the fly boys. 'Heelo 9er this is Alpha 9er, we have local troops three miles from extraction over.' 'Heelo 9er copy that. I confirm extraction. Out.'

All your teams arrive at the extraction point together although two men from Alpha 5 are being carried. You look to your sergeant, 'Form a perimeter, strong on that side please'. The men are already rushing to form a loose circle to protect the Blackhawks as they land

snipers to shoot in the plan and one in reserve. Shooting the bombers outside the house would minimize the chances of collateral damage to civilians in the house, and perhaps neighbours, from the detonation of further explosives which would be likely to occur in the event the house was stormed.

The sniper teams were armed with the bolt-action Accuracy International Inc. Arctic Warfare Magnum rifle which fires the .338 Lapua Magnum (8.58mm x 70mm) round. This is a specialized rimless bottlenecked centre-fire cartridge developed for military long-range sniper rifles. And it knocks down targets like they have been hit by a train.

On the morning of the attack the sniper teams took up position unnoticed and a US drone flew overhead providing live video coverage to the operation commander. Intelligence operators provided a live translation of conversations within the house.

At 08:00 the three terrorist suicide bombers left the house and the operation commander gave the order to open fire when they were well clear and all in vision at the same time. With three shots the bombers were killed, without detonating their explosive vests, and the support troops moved in to secure the area and search the bodies and building.

That is how an operation should work. Clean and quick with no drama.

A HH-60G Pavehawk loads men immediately after a successful operation. (USAF)

except for the wounded who are being tended in the centre. The choppers will come down one at a time to take up the men, wounded first. Moments later the distinctive sound of chopper blades brings a smile to a man's face. 'The cavalry's here.'

The first Blackhawk touches down, loads men and soars away. The next lands, loads and away, then the third and last approaches. A rocket trail lances up from behind a low hill and an explosion rocks the second chopper. Trailing smoke it switches to auto rotate and begins to lose height quickly. 'S**t!' The third chopper aborts its approach and speeds off low over the ground sprouting anti-missile flares as it goes. The damaged aircraft hits the ground with a crunch of metal just audible.

You call the Blackhawk commander: 'Heelo 9er, Alpha 9er do you have a visual on the downed chopper?' 'Heelo 9er the air frame in intact and men are climbing out over.' 'Alpha 9er can you extract them and us over?' 'Heelo 9er that is a negative. The locals have missiles. Sorry Buddy, withdrawing my callsign from range. Will remain within theatre, out.' The men close to you are watching. They know the choppers will leave them and their mates in the downed chopper rather than risk more aircraft and crew. What are you going to do?

You switch channels to your commanding officer. 'Sunray Alpha this is Alpha 9er over.'

'Sunray Alpha send over.'

'Primary mission accomplished. We have contact with local forces. They have missiles and one chopper is down. Ground team under fire. Request permission to destroy locals over.'

'Sunray Alpha wait out.' So now you just have to wait for a decision from brigade headquarters about destroying the missile launchers.

## Command decision one

*Air support is granted.*
As you wait first one, then a string of 82mm mortar bombs drop around your position. A heavy machine gun is firing high and you can see the tracer rounds flash over. The enemy are less than half a mile away and you have no heavy weapons. How long will HQ be in making a decision? Will someone think it is easier to make no decision than the wrong decision?

You make a decision and switch channels. 'Reaper 9er, Alpha 9er take down the enemy missile launchers and infantry over.' 'Reaper 9er confirm target missile launchers and infantry over.' You confirm, 'Alpha 9er confirm target missile launchers and infantry over.' 'Reaper 9er wait out.' Your sergeant looks you in the eye, 'He doesn't think you have clearance Boss. He's going to check.'

Then all hell breaks loose as missiles, Paveway bombs, cluster bombs, all sorts of munitions rain down on the enemy and destroy them in seconds. 'I think we owe the Reaper detachment a few beers lads.' The radio squawks into life, 'Alpha 9er this is Heelo 9er confirm continue extraction over.'

'Alpha 9er confirm extraction out.'

A Blackhawk swoops in to collect the last of your men while Heelo 9er collects the survivors and bodies from the stricken chopper himself. As you pull away a CH-53E Super Stallion, which has been held in reserve to allow for accidents, prepares to recover the Blackhawk airframe.

## Command decision two

*Alternate command decision when the air support is refused.*
'Alpha 9er this is Reaper 9er we do not have clearance for this mission over.'

You reply, 'Alpha 9er copy out.' You change channels and call the Pavehawk commander.

'Alpha 9er. Can you give me some suppressive fire with those miniguns over?'

'Heelo 9er. Sorry buddy, just had a straight negative order on that. Over.'

You think for a moment.

'Alpha 9er. You have 16 of my men aboard two of your callsigns. Can you drop the 14 who can walk for me?'

'Heelo 9er. I certainly can bud. So long as I can keep my ships away from those damn missiles. Will take figures 5 as they are on their way home. Where do you want them?'

'Alpha 9er. Just shy of the top of that hill a click north east of me on the side away from the enemy.'

'Heelo 9er, that is affirmative. They have a pair of grenade machine guns on board ready for infantry use. Shall I drop them as well? Do you want the enemy radio net jamming?'

'Alpha 9er both would be very welcome. Alpha 9er out.'

You turn to your sergeant, 'Billy, we are going to line out to face the enemy. They will either turn or fight. Lets see what they do.'

'Right Boss. I'll sort it.'

Billy scurries off to the men, bent low. As he leaves them each group moves to their new position facing the approaching enemy.

You change channels to address the men on the local net. 'All callsigns this is Alpha 9er. I want the enemy to stop before they reach us, and hopefully go home, so we will open a steady suppressing fire as soon as you see targets.

The first burst of gunfire comes from your line, quickly followed by another and then general firing breaks out. You take out your binoculars and scan the advancing enemy. There are more than 100 infantry approaching in a rough line and you can see three armoured vehicles in support. Eight-wheeler BTR 80 APCs firing 14.5mm heavy machine guns from their turrets. But the fire is still high. The enemy infantry drop to the ground 300 yards away and continue their approach in pairs, using the fire and movement system, with one man moving as his buddy gives suppressing fire. The mortar fire continues steadily and the heavy machine guns from the armour are getting the range. They have been joined by medium machine guns and rifle fire from the front of each vehicle.

'Looks like he is going to try for a win Billy. Getting sticky.'

You change channels and call the Pavehawk commander, 'Alpha 9er. Can you give me an ETA for touch-down over?'

'Heelo 9er. ETA minutes 1 over.'

The reinforcements confirm this over the comms. 'Alpha 4. We are on the ground 50m short of the top of the hill. The ship with Alpha 3 is landing now. The Pavehawk crew gave us a shiny new Golf Mike Golf and a crate of ammo over.'

'Alpha 9er. Get yourselves to the top of that hill and tell me what you see. Get the GMGs rigged up ASAP.'

The enemy are now within 200 yards and your rifle fire is bouncing off the BTR 80s but you have killed several of the infantry although you have three casualties of your own.

'Grenade launchers Billy.' Billy whistles loudly and signals for the men to begin using the grenade launchers slung under their rifles with a tap of his own.

'Will we be fixing bayonets then Boss?' Billy grins without humour.

You radio begins to speak, 'Alpha 9er this is Alpha 4 over.'

You reply instantly, 'Alpha 9er send over.'

'Alpha 4. My callsign and Alpha 3 are on top of the hill and can see about a hundred infantry skirmishing towards your position. There are three BTR 80s in close support and three more further back. There's a huddle of trucks and a mortar group with three tubes about half a click away. Over.'

'Alpha 9er. Stop those BTRs with the GMGs as soon as you like. Fire on the infantry with small arms and personal grenade launchers. Over.'

'Alpha 1. Copy out.'

Moments later the slow bursts of automatic fire from the GMGs starts up on the hill. First one then a second, then the third of the closer BTRs bursts into flame. Firing down onto the enemy from the higher position is like shooting fish in a barrel. The infantry advance stalls as the men become reluctant to move forward and their fire lessens as their casualties grow.

'Alpha 9er, Alpha 4, Switch fire to the distant BTRs, trucks and mortar group over.'

'Alpha 4 copy out.'

The GMG fire pauses then continues. The mortar fire stops. The infantry begin to retreat at the run, many of them brought down by machine gun and rifle fire from the hill. You call out to your men, 'Cease fire!'

A Pavehawk swoops in to begin collecting your men, now formed into a protective circle, while Heelo 9er collects the survivors and bodies from the stricken chopper himself. As you pull away a CH-53E Super Stallion, which has been held in reserve to allow for accidents, prepares to recover the smashed Pavehawk airframe.

# GLOSSARY

| | |
|---|---|
| **AP** | anti-personnel |
| **BSA** | Broad Spectrum Antibiotic |
| **civvy** | Civilian/Non-military personnel |
| **Cockneys** | Londoners, particularly working-class and from the East End of that city |
| **Det Cord** | detonation cord used to fire strings of charges |
| **False Flag** | Old naval term now used to describe an operation which is made to look like it has been perpetrated by another country's armed forces |
| **FOB** | Forward Operating Base |
| **GPMG** | General Purpose Machine Gun |
| **GMG** | grenade machine gun |
| **HEAT** | High Explosive Anti-Tank |
| **HAHO** | High Altitude High Opening |
| **IED** | Improvised Explosive Device |
| **IMT** | Individual Movement Technique |
| **MEMS** | micro-electromechanical systems |
| **NATO** | North Atlantic Treaty Organization (also known as the North Atlantic Alliance) |
| **Para(s)** | Parachute Regiment or members thereof |
| **PE** | plastic explosive |
| **PIAT** | Projector, Infantry, Anti-Tank |
| **PMC** | Private Military Contractor |
| **R&R** | rest and recuperation |
| **RPG** | Rocket Propelled Grenade |
| **Sangar** | a permannt, fortified, room-sized defensive or observation position |
| **SAS** | Special Air Service |
| **SF** | Special Forces |
| **SOPs** | Standard Operating Procedures |
| **stag** | period of sentry duty |
| **Stick** | a unit of 4 to 8 infantry or SF |
| **TAB** | Tactical Advance into Battle, from the Parachute Regiment concept of landing by parachute some distance from the objective and running to the battle |
| **TNBI** | neutron backscatter imaging |
| **Two-way range** | combat situation (ie. people are returning fire at you with live ammunition) |
| **UAV** | Unmanned Aerial Vehicle |
| **VOD** | velocity of detonation used to define the power of an explosive |
| **VP** | Voice Procedure, the accepted way to converse over radio |